Industrial Cogeneration Applications

Industrial Cogeneration Applications

Dilip R. Limaye

Published by
THE FAIRMONT PRESS, INC.
700 Indian Trail
Lilburn, GA 30247

Library of Congress Cataloging-in-Publication Data:

Industrial cogeneration applications.

"Extracted from papers presented at the Industrial Energy Technology Conference, sponsored by the Texas Public Utilities Commission (PUC) and held in Houston, Texas, in April 1985"–CIP acknol.

Includes bibliographies and index.

1. Cogeneration of electric power and heat–Congresses. I. Limaye, Dilip R. II. Industrial Energy Technology Conference (1985 : Houston, Tex.) III. Texas. Public Utility Commission.

TK1041.I534 1986 333.79'3 85-45873

ISBN 0-88173-022-X

Industrial Cogeneration Applications

Published by The Fairmont Press, Inc.
700 Indian Trail
Lilburn, GA 30247

ISBN 0-88173-022-X FP

ISBN 0-13-459249-2 PH

Printed in the United States of America.

Distributed by Prentice-Hall, Inc.
A division of Simon & Schuster
Englewood Cliffs, NJ 07632

Prentice-Hall International (UK) Limited, *London*
Prentice-Hall of Australia Pty. Limited, *Sydney*
Editora Prentice-Hall do Brasil, Ltda., *Rio de Janeiro*
Prentice-Hall Canada Inc., *Toronto*
Prentice-Hall Hispanoamericana, S.A., *Mexico*
Prentice-Hall of India Private Limited, *New Delhi*
Prentice-Hall of Japan, Inc., *Tokyo*
Prentice-Hall of Southeast Asia Pte. Ltd., *Singapore*

Acknowledgment

Most of the material in this book is extracted from papers presented at the Industrial Energy Technology Conference, sponsored by the Texas Public Utilities Commission (PUC) and held in Houston, Texas, in April 1985. The assistance and cooperation of the Texas PUC and the Conference Program Chairman, Dr. Milt Williams, are gratefully acknowledged. Special thanks are also due to the individual authors for their contributions, to Eileen Watts for editing the manuscript, and to Kim Coady and Lisa Cheever for the manuscript preparation.

Contributors

D.H. Cooke – Stone & Webster Engineering Corp., Houston, TX
Dr. Terry A. Ferrar – ANR Venture Management Company, Detroit, MI
Jack Flynn – Utility Savings Unlimited, San Diego, CA
Alfred B. Focke – Boyce Engineering International, Inc., Houston, TX
John J. Guide, P.E. – Exxon Chemical Co., Florham Park, NJ
William Hanzel – Nalco Chemical Co., Naperville, IL
Kenneth W. Jacobs, P.E. – Stanley Consultants, Muscatine, IA
Edward J. Landry – Pyropower Corporation, San Diego, CA
Don H. Lark – Energetics Systems, San Diego, CA
Michael K. Malloy – GEC Rolls-Royce, Houston, TX
R.H. McCue – Stone & Webster Engineering Corp., Houston, TX
Cyrus B. Meher-Homji – Boyce Engineering International Inc., Houston, TX
Roger W. Michalowski – Thermo Electron Corp., Waltham, MA
Gabor Miskolczy – Thermo Electron Corp., Waltham, MA
Dean Morgan – Thermo Electron Corp., Waltham, MA
Martin A. Mozzo, Jr., P.E. – American Standard Inc., New York, NY
David Pasquinelli – John Zink Company, Naperville, IL
Michael P. Polsky – Indeck Energy Services, Inc., Wheeling, IL
Norm B. Smith, P.E. – Stanley Consultants, Muscatine, IA
Dr. Adil Talaysum – Stuart School of Business, Illinois Institute of Technology, Chicago, IL
Cabot B. Thunem, P.E. – Stanley Consultants, Mucatine, IA
Roger Turner – Crawford & Russell, Inc., Stamford, CT
William L. Viar, P.E. – Waterland, Viar & Associates, Inc., Newark, DE
Joseph P. Wagner – Mechanical Technology, Inc., Latham, NY
Malcolm Williams – Gulf States Utilities Co., Beaumont, TX

Foreword

The cogeneration marketplace is growing and many new factors are influencing its future. A recent study conducted by the Cogeneration Institute confirmed that seventy-four percent of present cogeneration users are satisfied with system performance. In addition, these users planned to increase their capacity from two to ten fold in the next five years.

Incentives for utilizing cogeneration have also increased. "Electric rate shock" as a result of new nuclear plants coming on-line has greatly improved the economics for cogeneration in various parts of the country. This, coupled with stable or declining oil and gas prices, is giving cogeneration projects new impetus.

Protect financing is also becoming easier with several electric utilities establishing subsidiaries to help develop and finance cogeneration projects.

The industrial sector offers a great potential for cogeneration. Industries such as pulp and paper and chemicals have traditionally utilized cogeneration. Today there is much more opportunity to harness cogeneration's potential. It is hoped that this book will provide direction on how you can take advantage of cogeneration for your plant.

Albert Thumann, P.E., C.E.M.
Executive Director
Association of Energy Engineers

Industrial Cogeneration Applications
Dilip R. Limaye

Contents

PART I

COGENERATION TRENDS AND PROSPECTS

CHAPTER 1

Industrial Cogeneration: Trends and Prospects

Dilip R. Limaye

INTRODUCTION

Changing energy economics, combined with legislative and regulatory initiatives, have resulted in an increased emphasis on efficiency in industrial energy utilization. Cogeneration, the simultaneous production of electric (or mechanical) and thermal energy from a single energy source, has become very attractive for providing industrial energy needs economically and efficiently. The concept of on-site generation of electricity and the utilization of the thermal energy rejected in the power cycle for process energy needs, is certainly not new. According to the U.S. Office of Technology Assessment, over 59% of total U.S. electric generation capacity was located at industrial sites in the year 1900.[1] However, improved economics of electric power generation, transmission, and distribution, availability of inexpensive fuels, and regulatory barriers and constraints, all contributed to the decline of on-site generation and cogeneration.[2] By 1977, Industrial generation represented only 4% of total U.S. electricity generation.[3]

The significant changes in the availability and costs of energy, beginning with the 1973 Arab oil embargo, emphasized the need for efficient energy utilization, and led to the passage of the National Energy Act of 1978, which attempted to remove a number of economic and institutional barriers to increased use of cogeneration by industry. In particular, the passage of the Public Utility Regulatory Policies Act (PURPA) has led to a resurgence of interest in industrial cogeneration.[4] The implementation of the various provisions of PURPA by the Federal Energy Regulatory Commission and the various state regulatory commissions has taken longer than originally anticipated, partly because of the two major legal challenges to PURPA.[5] However, with the resolution of some of these implementation issues, the prospects for industrial cogeneration appear very attractive. This chapter provides an overview of the current status and prospects for industrial cogeneration.

CURRENT TRENDS

Changes in the economic and institutional factors influencing cogeneration that started in the late 1970s, are likely to continue into the 1990s. Industrial cogeneration is an idea whose time has come—again. The major legal challenges to PURPA have been resolved by the U.S. Supreme Court, and most of the states have developed their final rules implementing PURPA. Electric utilities, some of whom initially challenged PURPA and discouraged the increased implementation of cogeneration, are now increasingly tolerating, accepting, and in some cases, even welcoming cogeneration.[6] Many recent studies have pointed out the significant benefits of cogeneration, and the fact that cooperative efforts among industry, utilities and third parties can be mutually beneficial.[7]

The resolution of a number of legal issues related to PURPA and its implementation by states has led to what appears to be a more favorable and stable environment for cogeneration. A number of projects that were put on hold pending the outcome of the legal issues have now moved ahead to detailed planning, design and construction. The U.S. Department of Energy has recently published an optimistic projection of industrial cogeneration potential,[8] indicating a total market of 3,644 cogeneration projects with a combined capacity of 39,348 MW. The DOE projections are based on an assumption that cogeneration is economically feasible at a 7% return on investment (ROI), an assumption judged by many to be overly optimistic. These projections also appear to ignore some of the cogeneration capacity already on line. Nevertheless, the DOE study points out that there is significant cogeneration potential in the industrial sector. Similar studies at the state level, for example in Pennsylvania[9] and Illinois,[10] have also indicated the large potential for industrial cogeneration.

It is important to note, however, that the "old" cogeneration game is now played under new "rules," which determine what share of cogeneration-related benefits go to which various parties. This environment creates many opportunities for innovative business and institutional arrangements, and entrepreneurs who will recognize and take advantage of these opportunities will capture a greater share of the potential cogeneration benefits.[11]

The rapid changes of the last five years have led many organizations, from small third-party investment companies to industry giants such as General Electric, Combustion Engineering, and Babcock and Wilcox, to reshape their approaches and business plans for cogeneration. The marketplace is extremely dynamic. Many organizations traditionally involved in the design and construction of central station power generation plants, facing a significant decline in that market, are attempting to "retool" themselves and enter the cogeneration business.

There is considerable interest therefore in identifying the potential market and in determining the possible roles and niches of different organizations in it. Key issues that will influence the industrial cogeneration market are briefly discussed below.

KEY ISSUES AFFECTING THE FUTURE OF INDUSTRIAL COGENERATION

Regulatory Issues

In the past several years, two major legal challenges to PURPA were resolved by the U.S. Supreme Court. In the FERC vs. Mississippi case, the Court overturned a lower court ruling that PURPA was unconstitutional. In the American Electric Power lawsuit, the Court upheld FERC regulations regarding the requirement that utilities interconnect to cogenerators, and that 100% of avoided costs be used as the guideline for electricity sales by a qualifying facility (QF) to a utility.

As a result of these two decisions, much uncertainty relative to PURPA implementation at the Federal level was eliminated, and many states that had been waiting for the Supreme Court decisions moved ahead with PURPA implementation. The more interesting regulatory issues have now shifted to the state arena.

In New York, the State Public Service Commission (PSC) established a floor price of 6¢/kWh for electricity sales from a QF to a utility. This was challenged by Con Edison, who claimed that the floor price was higher than avoided costs and therefore was in violation of PURPA. A lower court upheld Con Edison's position, but, upon appeal by the PSC, the New York Supreme Court ruled that the floor price established by PSC was within the intent and rules established by PURPA.[12] It is widely believed that the New York ruling on floor price will influence a number of state regulatory commissions to take a similar stance on floor prices for cogeneration and will spur cogeneration installations.

At the same time that New York was ruling in favor of a floor price, the Kansas Supreme Court struck down buyback rates established by the Kansas Corporation Commission because these rates were higher than avoided costs. While the Kansas ruling is exactly the opposite of New York, most legal and regulatory experts believe that there will be a trend towards floor prices in excess of avoided costs for the early years of project operation.[13]

Other interesting regulatory developments with respect to state implementation of cogeneration include the establishment of separate avoided costs and rates for every 50 MW of incremental cogeneration in Maine, requirement for capacity pre-payments to municipal solid waste (MSW) cogenerators in Florida, and moves towards establishment of wheeling requirements for cogenerated power in some states.

A review of state implementation of PURPA indicates that states have taken widely divergent approaches to formulating rules and regulations relative to calculating energy and capacity credits, establishing standard offers, requirements for review and approval of negotiated contracts between utilities and cogenerators, etc.[14] The general trends in state regulation indicate that state commissions are becoming increasingly knowledgeable regarding major issues relative to cogeneration, and state regulatory postures are changing. While some commissions are becoming more favorably disposed towards cogenerators, many others are increasingly concerned about consumer impacts of cogeneration.

Economic Issues

While states are taking a favorable attitude towards cogeneration regulations, the reductions in fuel prices are leading to decreased avoided costs, and therefore less attractive buyback rates. These trends have also been heavily influenced by electric load growths that have been lower than anticipated, leading to deferrals or cancellations of many new power plants. Electric utilities in many parts of the country have successfully agreed that, since they have excess capacity and are not planning to add new capacity for a long time, there is no "avoided capacity" and therefore no capacity credit.[15] At the same time, utilities in other parts of the country have recognized that industrial cogeneration may provide a means for capacity addition without the need for the utility to generate additional financing. These utilities offer levelized rates for purchase of cogenerated power that increase the attractiveness of the cogeneration project in the early years.[16]

In general, however, it is becoming less attractive to design and build industrial cogeneration plants that are "oversized" and are able to export electricity to the local utility. The trend is towards systems that are matched to the on-site thermal load and are designed to displace purchased electricity. Some of the current controversy relates to issues such as retail sales of electricity and standby power for third-party-owned cogeneration projects.

Environmental Issues

A key event with significant negative implications for cogeneration was the ruling by the South Coast Air Quality Management District (SCAQMD) in California[17] to require gas turbine cogeneration plants to install selective catalytic reduction (SCR). This technology is not currently in widespread use in the U.S. The SCAQMD, however, based on limited experience in Japan with this technology, ruled that it was the best available control technology (BACT), and therefore must be installed on all gas turbine plants. The lack of this technology's general availability, the significant increased costs, and the need for careful monitoring and control, have substantial detrimental effects on gas turbine cogeneration economics. In the words of a California cogeneration developer, this ruling "has caused cogeneration in California to virtually come to a halt."[18] Other air quailty management districts in California are likely to follow SCAQMD's lead in this matter. Air quality concerns relative to cogeneration also exist in New York, Texas and Connecticut.

It remains to be seen whether air pollution control authorities in other states adopt the SCR requirements, and what the actual impact of SCR will be on the economics of cogeneration. It appears at this time, however, that this ruling by SCAQMD is a major blow to the future development of cogeneration.

Tax Laws

The recent Treasury proposals to make widespread changes in tax laws also could have a detrimental impact on cogeneration.[19] In particular, the proposal to change from the Accelerated Cost Recovery System (ACRS) to the Real Cost Recovery System (RCRS) could remove some of the significant depreciation benefits associated with qualifying facilities under PURPA. Substantial revisions to the Treasury proposals are likely, and the actual changes in tax laws are unclear. With the administration pushing for tax reform and increased tax revenues, it is possible that some benefits currently available to cogenerators may disappear. This may have a negative effect on the availability of third-party financing for cogeneration projects.

These changes in tax and depreciation laws are likely to eliminate marginally economic industrial cogeneration projects that were attractive only because innovative financial and organizational structures took advantage of the tax and depreciation benefits. In the future, the emphasis will be on projects that are inherently attractive because of the efficiency and economics of cogeneration.

New Technologies

While cogeneration is not a new concept and some cogeneration facilities have been in operation for more than 50-60 years, there are several new technological developments that could be beneficial to cogeneration. Some of the new technologies include:

- *Steam Injection Cycle,* in which steam is injected into the gas turbine to increase power output. This cycle may be economically attractive for facilities with widely fluctuating thermal and electric loads. When thermal needs are low, steam injection provides additional power for the facility or for export to the utility. This technology may satisfy air quality requirements without SCR.
- *Kalina cycle,* a dual-fluid cycle which uses an ammonia-water mixture as the working fluid. The cycle offers increased output from a steam turbine.

- *Fluidized-Bed Air Turbine,* which uses the hot gases from combustion to drive a gas turbine. This technology would make solid fuels (such as coal, petroleum coke, peat, wood chips, wood waste, etc.) feasible for use with combustion turbines, thereby making cogeneration more attractive in a number of industries where such fuels are readily available.
- *Fuel Cell,* which uses an electro-chemical process with a high electric efficiency. Among the fuel cell's many advantages are its modularity, flexibility and ability to provide varying thermal loads without a degradation in its electric output.

Furthur development and commercial availability of these technologies is likely to increase the economic attractiveness of cogeneration.

ROLES OF INDUSTRY, UTILITIES AND THIRD PARTIES IN COGENERATION IMPLEMENTATION

Industry Role

Despite the proven economic attractiveness of cogeneration and the significant publicity it has received in the past several years, many industrial managers are reluctant to invest in cogeneration. There are many reasons for this reluctance:

- Capital availability is always a major issue and investments more closely related to the basic business generally have a higher priority than cogeneration.
- Industrial executives generally want a high return on investment. Many cogeneration projects, while economically attractive, do not satisfy industry's hurdle rate requirements.
- In industries where on-site power generation is not common, there is a lack of skills and expertise needed to operate a power boiler or a combustion turbine.

- Despite the existence of cogeneration for many decades, some industrial managers have a perception of potential risk in investing in this option.
- Industrial managers view their business as manufacturing certain products—not electricity generation.

Many of these barriers are being overcome through joint ventures with third parties or with utilities. Also, as the benefits of cogeneration are demonstrated through the implementation of many projects, industrial executives are expressing greater interest in this option.

Role of Third-Party Financing

Some of the financial barriers to cogeneration implementation have been overcome by third-party financing organizations. The availability of such financing has allowed many industrial firms to implement cogeneration projects which may have had marginal economics (from the viewpoint of industrial hurdle rate requirements) or may have faced capital availability constraints. Third-party financing organizations can take advantage of tax and depreciation benefits and are often willing to accept rates of return that appear marginal from the industrial perspective. Third-party financing has therefore been very popular in the field of cogeneration.

Third-party financing organizations have ranged from small entrepreneurs to large organizations such as General Electric Credit Corporation.

Role of Utilities

Electric utilities face unprecedented challenges resulting from the changing economics of power generation. In particular, the very high costs of generating capacity additions and the resulting financial risks have deterred many utilities from building new major generating facilities. Consequently, there is a definite trend towards increased interest in and acceptance of industrial cogeneration by utilities. Many are now including industrial cogeneration in their future capacity planning, and some have recognized industrial cogeneration as an alternative new business opportunity. Cogeneration projects, partially owned and operated by a utility, can be highly complementary to the utility's traditional business and may provide opportunities for unregulated earnings.

Many utilities are examining opportunities for participation in industrial cogeneration ventures, and some have set up subsidiaries specifically for such participation.[20]

It appears that the changing economic and institutional environment will lead electric utilities to gradually redefine their traditional role. In the future, utilities are likely to be seen as "energy service companies" rather than merely as suppliers of electricity. In this new role, utilities may embark upon many new types of business ventures, some of which have already been undertaken by utilities in the past several years. Thus, utilities may actively encourage cogeneration and may even participate in such projects.

THE FULL-SERVICE COGENERATION BUSINESS

Recent developments in the cogeneration market have attracted a number of organizations into the "full service cogeneration business." Full-service organizations (or consortia of organizations) can provide a complete range of services to get a cogeneration project on line—from prefeasibility and feasibility analysis to design, engineering, construction, financing, operation and maintenance, and equity ownership. All the industrial client need do is sign contracts for the purchase of steam and electricity (or steam only, with electricity sold under contract to the local utility).

The types of organizations interested in offering full services to potential cogenerators include:

- Equipment manufacturers/suppliers
- Architect/engineering firms
- Financial organizations
- Energy management companies
- Entrepreneurs/developers
- Fuel suppliers.

Many of these organizations, facing significant changes in their traditional markets, view cogeneration as an attractive and growing market and are committing significant resources in attempts to obtain a share of this market. During the last several years, a number of "industry giants" have announced their entry into the full-service cogeneration business. Most of these organizations are willing to make significant equity commitments to cogeneration projects.

IMPLEMENTING COGENERATION PROJECTS–SOLVING THE RUBIK'S CUBE

The implementation of cogeneration projects has been likened to the solution of the Rubik's cube.[21] This famous puzzle involves manipulating a six-colored cube in such a way as to obtain one color on each of the cube's six faces. The complexity of the problem stems from the fact that a change made in any one face influences a number of other faces.

The six faces of the "cogeneration cube" are (see Figure 1-1):

- technology/systems
- fuels
- economics
- financing
- regulation
- contract negotiation

Attempts to optimize the project by adjusting any of these elements invariably influence many others. For example, the choice of a technology (or type of system) influences the fuel choice and economics. Financing methods influence economics and contract negotiation, and regulatory considerations influence all other elements.

Solving this cogeneration puzzle requires a great deal of skill, experience and ingenuity. Many organizations have found the complexities overwhelming. The entry of the full-service organizations, who have substantial experience with a number of these elements, will facilitate appropriate treatment of the numerous concerns in these six areas in order to implement projects successfully. However, even some industry giants do not possess all the skills and capabilities required to address cogeneration's many facets. Consequently, a significant amount of joint venturing of organizations occurs to implement projects, to solve development problems, and to share risks.

Figure 1-1
THE COGENERATION CUBE

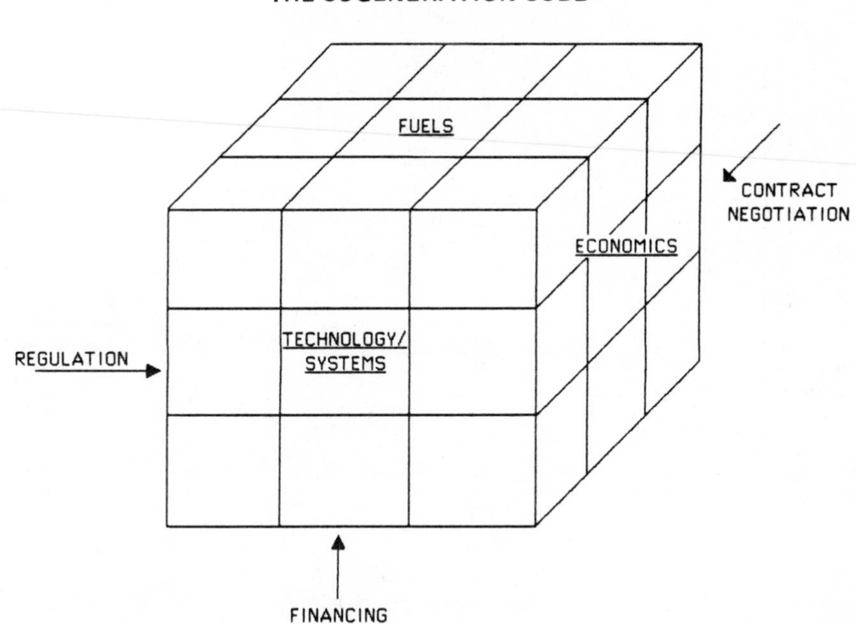

PROJECTIONS OF INDUSTRIAL COGENERATION

Despite the uncertainties relative to some factors influencing industrial cogeneration, there is general agreement that its prospects are excellent. Figure 1-2 shows projections of total electric generating capacity in the U.S. at three different growth rates–low (1% per year), medium (2%) and high (3%). During the past five to six years, approximately 15 to 20% of the generating capacity additions have come through industrial cogeneration.[22] Assuming that this proportion applies in the future, the range of cogeneration capacity additions is shown in Figure 1-3.

Table 1-1 shows the prospects for cogeneration by major industry groups. The most attractive industries are likely to be chemical, petroleum, pulp and paper, and food. Table 1-1 also shows the most likely system and fuel type, as well as the size of typical systems.

Figure 1-2
PROJECTIONS OF ELECTRIC GENERATION CAPACITY

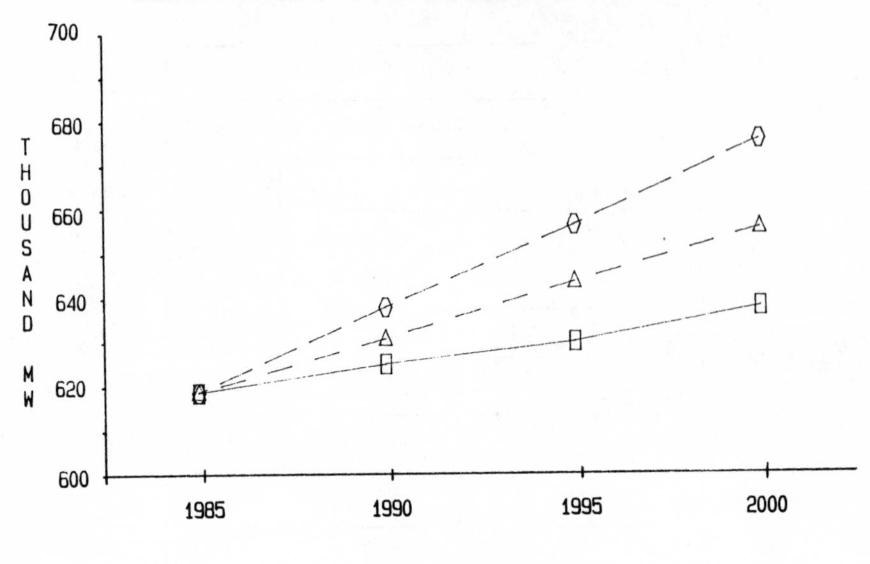

Figure 1-3
PROJECTIONS OF COGENERATION CAPACITY

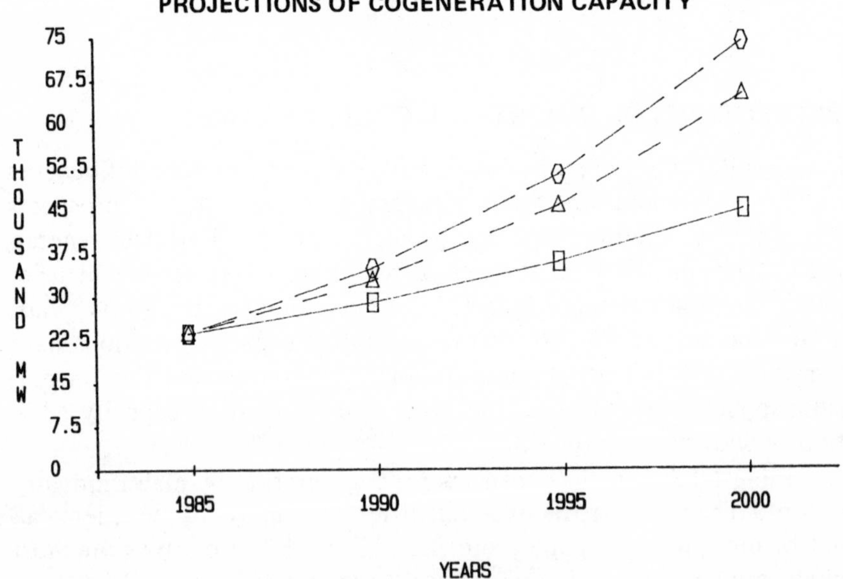

Table 1-1

INDUSTRIAL COGENERATION PROSPECTS
(1986-2000)

	Cogeneration Potential 1986-2000				
Industry Type	**No. of Projects**	**Typical MW Size**	**Total MW Capacity**	**Most Likely Cycles/Systems***	**Most Likely Fuels****
Pulp & Paper	30-50	15-100	1500-3000	ST, FBC	B, C
Petroleum	30-60	10-100	1000-3000	ST, GT, CC, FBC	BP, NG
Chemical	40-100	10-150	4000-8000	GT, CC	NG, C
Food	50-150	3-50	1000-3000	GT, ST	BP, NG
Textile	50-100	2-20	500-1000	ST, GT	NG
Rubber/Plastics	10-50	2-25	200-500	GT, CC	NG, C, O
Steel	0-25	5-40	0-400	ST, GT, B	C, NG, WH
Lumber/Wood	50-150	2-25	400-1000	ST, FBC	B
Cement/Glass	10-50	2-25	200-500	GT, B	NG, WH
Metal Fabrication	20-100	5-40	200-2000	GT, CC	NG
Industrial Parks	5-20	50-200	500-3000	CC, GT, ST	C, B, NG

*ST = Steam Turbine, FBC = Fluidized Bed Combustion, GT = Gas Turbine, CC = Combined Cycle
B = Bottoming Cycle

**B = Biomass, C = Coal, O = Oil, NG = Natural Gas, WH = Waste Heat, BP = By-product Fuel

COGENERATION TRENDS AND PROSPECTS

Based on the considerations listed above, the current status and future prospects of the industrial cogeneration market can be characterized as follows:

- The potential market is large. Even if we conservatively estimate that over the next 15 years (1985-2000) about 20,000 MW of industrial cogeneration will be installed, the total market exceeds 25 billion dollars.
- The marketplace will continue to be very dynamic.
- The market will be influenced strongly by a number of regulatory, technical, economic and environmental issues.
- The competition in the full-service business will continue to increase.
- The size of the projects is decreasing. Future cogeneration projects are likely to be sized to meet facility electric loads (i.e., displace electricity) rather than to export large quantities of electricity to the local utility.
- Many important skills need to be developed or assembled for successful implementation. The entrepreneurial role of the project "developer," in lining up the project and the required skills, and in "deal-making," is a critical factor in the successful implementation of cogeneration. Small organizations with such entrepreneurial skills (e.g., Power Systems Engineering, Applied Energy Services, Decker Energy International) have played a significant role in the market.
- As new entrants in the market develop/assemble the needed skills, a significant number of joint ventures are being formed. The joint ventures are all project-specific, and many different types of ventures have been developed.

The industrial managers with the "right type of facility" are likely to find a number of organizations selling them the full-service concept. They should be careful about selecting an organization for such a full-service project, because few organizations have yet assembled the needed skills and experience to successfully solve the cogeneration cube.

The efficiency of cogeneration systems in meeting industrial energy needs makes them economically attractive. Cooperation among industry, utilities and third-party developers/financiers will lead to the implementation of optimum cogeneration systems with significant economic and environmental benefits to industry, utility and society.

ACKNOWLEDGEMENT

Portions of this chapter have been adapted from an earlier article by the author titled, "A Strategic View of the Cogeneration Market," published in *Strategic Planning and Energy Management,* Vol. 4, No. 4, Spring 1985.

REFERENCES

[1] Office of Technology Assessment, *Industrial and Commercial Cogeneration,* OTA-E-192, Washington, DC, February 1983.

[2] Synergic Resources Corporation, *Evaluation of Dual Energy Use Systems,* EPRI EM-2695, Electric Power Research Institute, Palo Alto, CA, 1982.

[3] Frederick H. Pickel, *Cogeneration in the U.S.: An Economic and Technical Analysis,* M.I.T. Energy Laboratory Report, MIT-EL-78-039. Boston, MA 1978.

[4] Thomas C. Hough and Dilip R. Limaye, *Utility Participation in DEUS Projects: Regulatory and Financial Aspects,* SRC Report 7070-R1, Bala Cynwyd, PA, 1981.

[5] Dilip R. Limaye, "A Strategic View of the Cogeneration Market," *Strategic Planning and Energy Management,* Vol. 4, No. 4, Spring 1985.

[6] Synergic Resources Corporation, *Proceedings of the Utility Cogeneration Symposium,* Electric Power Research Institute, Palo Alto, CA, 1986.

[7] Dilip R. Limaye, "Evaluation of Industry/Utility Cooperative Efforts in Cogeneration," Paper presented at the Fifth World Energy Engineering Congress, Atlanta, GA, September 1982.

[8] U.S. Department of Energy, *Industrial Cogeneration Potential for Application of Four Commercially Available Prime Movers at the Plant Site,* Washington, D.C., August 1984.

[9]Synergic Resources Corporation, *Assessment of Industrial Cogeneration Potential in Pennsylvania,* Final Report, submitted to the Governor's Energy Council, 1983.

[10]Synergic Resources Corporation, *An Evaluation of the Technical Potential for Cogeneration in Illinois,* Final Report, submitted to Illinois Department of Energy and Natural Resources, 1985.

[11]Dilip R. Limaye, "Cogeneration: Trends and Prospects." *Cogeneration Journal,* Vol. I, No. 1, Fall 1985.

[12]"Six Cents/kWh Upheld by New York," *The Cogeneration Letter,* December 1984, p. 5.

[13]Personal communications, Michael Zimmer, The Cogeneration Coalition, and Lee Goodwin, Wickwire, Gavin and Gibbs.

[14]Dilip R. Limaye, "Overview of State Regulation of Cogeneration," Paper presented at the Workshop on Cogeneration: Federal/State Regulation, Arlington, VA, April 1983.

[15]*Energy User News,* "States' Cogeneration Rate-Setting Under PURPA," various issues, 1985.

[16]"Massachusetts Municipal Agency Eyes Program to Buy From Small Power Producers," *The Cogeneration Report,* July 19, 1985, p. 3.

[17]"Environmental Woes in California," *The Cogeneration Letter,* February 1985.

[18]Personal communications with a representative of a California cogeneration development company.

[19]Lee Goodwin, "Treasury Recommends Major Tax Changes for Cogeneration and Alternative Energy Industries," unpublished memo, November 30, 1984.

[20]"Eleven Utilities Now Have Cogeneration Subsidiaries," *Cogeneration Report,* January 18, 1985, p. 9.

[21]Thomas Casten, "Cogeneration/District Heating Implementation: A Developer's View," in D.R. Limaye (Ed.) *Planning Cogeneration Systems,* Atlanta: Fairmont Press, 1985.

[22]Information based on Cogeneration Data Base developed by Synergic Resources Corporation.

CHAPTER 2

The Dynamic Nature of Cogeneration Or "The PURPA Amoeba"

Michael P. Polsky and Edward J. Landry

INTRODUCTION

According to Webster, "Amoeba" is derived from the Greek word meaning "to change." "Amoeba" is defined to be "any of a large genus of naked rhizopod protozoans with lobed and never anastromosing pseudopodia and without permanent organelles. . ."– certainly a definition as understandable to the layman as the topic of cogeneration itself.

Over the years, we have seen substantial changes in the cogeneration industry. Although the PURPA legislation itself is not overly complicated, its implementation by the states has been a difficult and complex process. A major reason is that, although it is not a new technology, cogeneration's widespread development in competition with central station utility power plants is new. PURPA broke down barriers and obstacles to cogeneration development that had previously existed, upsetting an "equilibrium" situation in the utility industry. During the dynamic period that followed, certain trends were established as regulatory commissions, utilities, and cogenerators strove to reach a new equilibrium point. This chapter examines some of those trends and attempts to identify possible future ones.

Before discussing trends, it would be helpful to understand the basic fundamentals that govern the profitability of cogeneration projects. We will analyze cogeneration profits and explain why cogeneration economics may not always coincide with the maximum conservation of energy. We will also try to provide insight into cogeneration energy profitability for both the ultimate energy user and third-party developer.

ANATOMY OF COGENERATION ENERGY PROFITABILITY

Any true cogeneration cycle results in substantially higher fuel utilization efficiency than separate energy production. This is attributed to the fact that cogeneration is a sequential production of electrical and thermal energy, thus utilizing rejected energy from one process as useful energy input into another.

Let us formally examine cogeneration profits as a function of various efficiency and energy cost factors. Cogeneration profitability is measured as the difference between the costs of electrical and thermal energy for cogeneration and noncogeneration. (Noncogeneration energy is taken to be that produced by the utility and industrial steam boilers.) Because most of the energy cost savings in cogeneration is derived from a reduction in the amount of fuel consumed, relative energy profitability can be defined as the difference between the expenditures for fuel in cogeneration and the noncogeneration.

$$\Pi Q_{cq} = [(C_{fec}Q_{fec} + C_{fb}Q_b) - C_{cg}Q_{fcg}]/C_{fcg}Q_{fcg} \qquad (1)$$

If we replace the amounts of fuel required to produce a certain amount of useful energy, Q, with the amount of useful thermal and electrical (N) and (Q_t) energy and divide by the thermal efficiency of the specific energy source [refer to equations (2), (3), and (4) below], we can obtain the expression for the relative cogeneration profitability in terms of thermal efficiencies for the various energy sources [equation (5)].

$$Q_{fec} = N/\eta_{ec} \qquad (2)$$

$$Q_b = Q_t/\eta_b \qquad (3)$$

$$Q_{fcg} = (N + Q_t)/\eta_t \qquad (4)$$

$$\Pi Q_{cg} = \frac{C_{fec}}{Cfcg} \cdot \frac{\eta_{ecg}}{\eta_{ec}} + \frac{C_{fb}}{C_{fcg}} \cdot \frac{(\eta_t - \eta_{ecg})}{\eta_b} - 1 \qquad (5)$$

One important parameter in equation (5) is η_{ecg}

$$\eta_{ecg} = \frac{N\eta}{Q_{fcg}} = \frac{N \cdot \eta_t}{N + Q_t} \qquad (6)$$

η_{ecg} is the electrical efficiency of a cogeneration plant based on total fuel input. The principal characteristic of η_{ecg} is that in calculating it, no credit is given for useful thermal energy production. η_{ecg} characterizes the efficiency of a cogeneration plant in converting total fuel energy into electricity. As we will see later, η_{ecg} plays a very important role in determining overall cogeneration profitability.

If the cost of fuel for all the energy sources is the same,

$$C_{fec} = C_{fb} = C_{fcg}$$

then

$$\Pi Q_{cg} = \eta_{ecg}/\eta_{ec} + (\eta_t - \eta_{ecg})/\eta_b - 1 \tag{7}$$

Replacing η_{ecg} in equation (7) by the electrical heat rate, H.R. = $3413/\eta_{ecg}$, it follows that

$$\Pi Q_{cg} = \frac{(3413)}{\text{H.R.}}\left(\frac{1}{\eta_{ec}}\right) + \frac{\eta_t - 3413/\text{H.R.}}{\eta_b} - 1 \tag{8}$$

The first and second components in equations (7) and (8) are the absolute contributions of electrical energy and thermal energy, respectively, to cogeneration profits.

The second component in evaluations (7) and (8) is always greater than zero. If a cogeneration plant supplies no thermal energy at all, the $\eta_t = \eta_{ecg}$ and the contribution of thermal energy in cogeneration profits also becomes zero.

Also from equations (7) and (8) it follows that if the electrical efficiency of the cogeneration plant η_{ecg} is greater than the efficiency of the utility plant (η_{ec}), the cogeneration plant would provide profits *regardless of the supply of the thermal energy!*

In order to simplify further presentation, let us designate electrical and thermal components as

$$\frac{\eta_{ecg}}{\eta_{ec}} = \frac{(3413)}{\text{H.R.}} \cdot \left(\frac{1}{\eta_{ec}}\right) = A \tag{9}$$

$$\frac{\eta_t - \eta_{ecg}}{\eta_b} = \frac{\eta_t - 3413/\text{H.R.}}{\eta_b} = B \tag{10}$$

Then equation (8) becomes

$$\Pi Q_{cg} = A+B-1 = \frac{A(A+B-1)}{A+B} + \frac{B(A+B-1)}{A+B} \tag{11}$$

Equation (11) is equivalent to equation (8) but presents the overall profits as a sum of the weighted absolute contributions of the electrical and thermal energy in the overall profits. For example, the first and second components are the portions of the total profit (A+B−1) associated with the production of electrical and thermal energy respectively.

In order to analyze equation (11), let us assume that

electric utility efficiency η_{ec} = 0.335 (H.R. = 10,200 Btu/kWh),

industrial boiler efficiency η_b = 0.80, and

overall cogeneration efficiency η_t = 0.70.

Figure 2-1 is a graphical representation of equation (11) for various cogeneration net electrical efficiencies (η_{ecg}) or heat rates (H.R.).

From this figure we can draw some general conclusions:

1. Electricity contributes a much greater share of the overall cogeneration profit than thermal energy when electrical efficiencies are reasonably high.

2. If the cogeneration plant electrical efficiency is greater than the electrical efficiency of a utility plant, then the cogeneration plant is energy profitable regardless of whether thermal energy is produced or not.

3. Overall cogeneration profit, as well as its electrical component, decreases rapidly with a decrease in electrical efficiency or increase in heat rate. Higher electrical efficiency results in much greater fuel savings and profits even though overall thermal efficiency remains unchanged.

4. The contribution of thermal energy to overall cogeneration energy profit is relatively stable. Actually, it increases slightly with a decrease in electrical efficiency and then starts to decrease as elec-

trical efficiency continues to decline. This anomoly can be explained by the fact that despite increases in the production of thermal energy, decreases in total profits are overwhelming. However, thermal energy plays a greater role in overall cogeneration profits as electrical efficiency drops off (for example, steam turbine topping cycle).

HOW ENERGY PROFITS DIFFER FOR THE ENERGY USER AND THIRD-PARTY DEVELOPER

Now, we would like to evaluate various cogeneration plant sizing and equipment selection philosophies for the cases of the industrial owner (end-product user) and the third-party owner. There are two extremes in cogeneration plant sizing:

1. The plant is sized to produce only cogeneration electricity (a 100% cogeneration plant). In this case the size of the plant is determined by the thermal requirements.

2. The plant is sized to supply only a minimum amount of thermal energy (as required by PURPA).

Total energy profits, as defined above, for these two extremes are plotted in Figure 2-2. (Note that in order to simplify the presentation without sacrificing generality, we assume there will be no thermal production for the second case plant.)

For example, if the cogeneration plant were comprised of a General Electric Frame 7 combustion turbine* with a waste heat boiler, the total energy profit would be that represented by Point A on the 100% cogeneration line. If, at the other extreme, the plant were a combined cycle providing only electricity, the total profit would be as shown by Point B on the noncogeneration line. The line connecting Points A and B shows cogeneration profits for the plant producing various combinations of thermal and electrical energy.

As we can see from Figure 2-2, the 100% cogeneration plant would produce the maximum relative energy profit and the noncogeneration plant the minimum. If an industrial user were to own a cogeneration plant, his profits would be based on full avoidance

*G.E. turbines are mentioned merely to convey the characteristics of the machine used in this illustration. A similar turbine by another manufacturer would suffice equally well.

of noncogeneration thermal and electrical energy costs and would be on the line connecting the appropriate points on the 100% cogeneration and noncogeneration lines. The location of this point will be determined by the size of the plant.

Third-party ownership normally results in a thermal energy price (and possibly an electrical energy price) that is substantially discounted over the noncogeneration cost. As a consequence, the third party does not see all of the cogeneration profits that would be gained in the case of industrial ownership. Actually, it shares some of the profits (in the form of a discounted thermal energy price) with the end product user. For example, if the third party were to give a 50% thermal energy price discount, its total energy profits would be as indicated by Point C in Figure 2-2. If we connect Points B and C we can make some important observations:

1. With some energy price discounts (most often with thermal, but electrical price discounts can also apply), third-party energy profits may actually increase with an increase in electrical production by cogeneration. Therefore, it may be economical for a third party to install a larger cogeneration plant than dictated by thermal requirements.

2. The electrical efficiency of the plant can affect its sizing. For example, if we consider a cogeneration plant using either a high efficiency gas turbine such as a GE LM 2500 or a lower efficiency gas turbine such as a GE Frame 6, from Figure 2-2 it follows that a 100% cogeneration plant with the higher efficiency gas turbines would produce a substantially higher energy profit than the one with the units of lower efficiency (Points D and E respectively). Furthermore, we can install additional LM 2500 turbines to produce only electricity and almost double the plant size, and the energy profits will not be less than in the case of Frame 6 plant, Point F on Figure 2-2.

The result of the above is that in order to maximize its profits, the third-party developer would try to build the largest possible cogeneration plant within PURPA efficiency standards. The size of the plant could be expanded simply by installing a noncogeneration combined cycle condensing portion. Since the efficiency of a modern combined cycle is greater than the efficiency of an average utility plant, it would result in extra profits. We have already seen examples of such gains at several plants in the Gulf Coast Area and California.

NOTICEABLE COGENERATION TRENDS

To those who have followed cogeneration since its rebirth and through its formative years, certain trends can be identified. These trends have not necessarily appeared everywhere simultaneously, but in general, we have seen them arise in different areas of the country. Some of these trends have not yet occurred in some areas, but can be expected in the future.

Trend I: Changes in Utilities' Philosophies

The nature of cogeneration is very much a function of geography (state and region) as well as each electric utility's unique situation. The manner in which a utility perceives cogeneration depends upon a number of factors: reserve margin, current construction problems and capacity expansion plans, degree of industrial load (cogeneration potential), utility commission's position on the issue, fuel mix, etc.

Initially, utilities considered cogeneration as a threat to their monopoly of generating power to serve their customers. No longer would future load growth be met exclusively with utility-built power plants. This presented utilities with a scenario of stagnation—fewer (or no) new power plants in rate base and reduced revenues from lost sales. Understandably, utilities saw cogeneration as a real challenge to their business, and some observers feel they responded by fighting it through delay, non-cooperation, and outright unreasonableness in dealing with cogenerators.

After some time, however, utilities realized that such stalling tactics were delaying the inevitable. At the same time, increased pressure from regulatory commissions, who recognized the benefits of cogeneration, resulted in a reluctant change in attitude among utilities. Discussions with potential cogenerators became more open and some projects moved forward. Utilities were careful, however, to avoid opening the flood gates.

Some utilities ultimately recognized that one possible way to benefit from an otherwise undesirable situation was to own a portion of the project. As a result, utilities spun off subsidiaries to invest in such projects and, in some cases, attempted to encourage their development. Full-scale encouragement of cogeneration, however, will never occur as long as present laws and regulations restricting utilities' participation remain.

Trend II: State Public Utility Service Commissions Establishing Ground Rules

Cogeneration got off to a slow, shaky start while state commissions wrestled with the question of how to implement PURPA. Most states adopted rules that merely echoed the PURPA language and did little to answer such critical questions as, "How does one calculate avoided cost?" California certainly led the country in implementing a clear, workable policy, and the magnitude of Qualified Facility (QF) development there testifies to that fact.

State commissions are now going through a "second round" of implementing PURPA. Issues such as avoided cost methodologies, wheeling, queuing for available capacity needs, retail sales, etc., are now being addressed. After this second iteration of rule making is complete, everyone should have a better understanding of the ground rules. Project sponsors will then know fairly early whether their projects are viable.

Trend III: Growth of Third-Party Cogenerators

As interest in cogeneration exploded shortly after the passage of PURPA, the economy refused to cooperate. Interest rates soared and a recession dampened any outlook for industrial expansion. At the same time, however, energy prices were rising and cogeneration still seemed a logical course of action. New tax laws that made investment in such projects attractive also played an important role in maintaining industry's interest in cogeneration.

During this time period, companies that were saddled with capital constraints turned toward the third-party own and possibly operate approach. It was a way to enjoy the benefits of cogeneration without capital outlay. Other companies found this approach attractive because of their inability to utilize tax benefits or because of their reluctance to "get into the utility business." For whatever reason, there was indeed (and still is) tremendous interest in this alternative, and there are more than enough willing developers/ owner-operators.

The significance of this trend is that QFs became "investors" rather than "thermal users." The goal therefore became "maximize return" rather than "reduce energy costs." As we shall see, there is a trend occurring which counter-balances this one.

Trend IV: Larger Cogeneration Plants

Because of the existence of Trend III, we have seen a companion trend toward larger facilities. As shown previously, in many cases, profit motivation dictates projects with larger electrical output, especially for gas-fired systems. Thus, the so-called "PURPA machine" was born, and these projects grew to the maximum size allowed under PURPA standards.

Addition of a large increment of cogeneration capacity lowers utility marginal energy costs as well as eliminates the need for additional capacity from other cogenerators. As a result, the capacity and energy payments are reduced to the level that the larger "PURPA machine" type of projects are no longer economically viable and even risky.

Trend V: Oversupply of Potential Cogeneration Capacity

As a result of the trends discussed above, we have understandably witnessed a tremendous growth in the number of QFs either actually completed or at some stage in the pipeline. With few exceptions, however, electric utilities currently have more than enough installed capacity to meet their electric needs. Many large coal and nuclear projects under construction will add to this capacity margin when they come on-line.

This over-supply of cogeneration capacity has recently been observed in places like California, Texas and Oklahoma. State regulatory commissions are now attempting to deal with the problem. Solutions such as queuing ("first-come, first-served") in allocating capacity payments have been discussed. Other solutions such as competitive bidding have been proposed. In any event, this too is a self-correcting trend in that, as the avoidable capacity increments of the utility are fully subscribed and as more cogenerators displace higher-cost utility power plants, avoided costs will decline.

Trend VI: Smaller Projects

As previously mentioned, the days of the "PURPA machines," in most cases, seem to be over. In fact, an opposite trend toward smaller projects displacing purchased electricity only is under way. The main force behind this phenomenon is the low avoided costs as well as economic and regulatory uncertainty associated with selling power to the utility.

Projects of this type, in addition to being smaller in the level of electrical output, are also more balanced in the ratio of thermal to electrical energy output, and directed toward maximizing of energy savings, not electrical output. This is another trend away from the "PURPA machines."

A greater percentage of the revenue/saving stream is from the thermal side, making their economics somewhat less susceptible to swings in electricity rates. In addition, smaller projects of this type will, in many cases, tend to be owned by the energy user because they are less capital intensive and because the user desires control over his energy source and does not want to be involved in the "utility business."

FUTURE TRENDS

What about future trends that have not yet developed? How will the "PURPA Amoeba" change shape in the future? This is difficult to predict, but certain events can be identified that will influence cogeneration. One can then speculate on what trends may develop as a result of these events.

The current Administration is under increased pressure to reduce the federal deficit. The long-discussed tax changes will probably occur within the next year or two. How wide sweeping the changes will be and what areas will be specifically impacted, are anyone's guess.

The consensus, however, seems to be that the investment tax credit will be repealed and the depreciation schedule lengthened. Corporate tax rates may also change. One can therefore speculate that projects will have to rely less on tax advantages to be economically justified. Consequently, we may see fewer marginal third-party projects. Those projects that are solidly based on energy savings will remain attractive.

Wheeling remains an area of uncertainty that will gradually become clearer as more states deal with the issue. In a case involving the State of Florida, the FERC ruled that it had exclusive authority to set wheeling rates if the transmission of electric power in question were in interstate commerce (which includes almost all grids). In the same case, the FERC refused to rule on the question of whether states have the authority to require wheeling.

Wheeling will be especially important in states and regions where utilities have a capacity need. Several states, including Pennsylvania, Florida, and Texas, have regulatory provisions which permit their commissions to require wheeling of cogenerated power by utilities. Others may follow suit, and a FERC decision on whether states have the authority to require wheeling will probably occur soon. Resolution of this issue in favor of cogenerators will facilitate development of projects.

Despite the apparent pressure to tighten cogeneration qualification requirements, several decisions made recently by the FERC have relaxed the standards in PURPA regulations designed to ensure that only facilities of a certain type and efficiency qualify.

However, it is possible that pressures by the utility lobby may finally result in changes to PURPA. Potential changes include tighter operating and efficiency standards for gas- and oil-fired cogeneration. Another possibility is the allowance for greater utility ownership in QF projects.

Each state is (or will be) adopting its own method of dealing with situations where there is more QF capacity potentially available than there is need. The State of Texas has decided to encourage competitive negotiations between utilities and QFs in such a situation in order to reduce the level of payments for electricity to as far below full avoided cost as possible. This, in essence, is a form of price competition.

Such an approach provides maximum economic benefits to the rate payers. The risks of insolvency to the "successful" project bidders because of low avoided cost payments will either be borne by the rate payers or, more likely, by the project sponsors through performance bonds or corporate guarantees. Although this system fails to equitably balance risks to rewards, it may be adopted in other states, particularly those whose regulatory commissions are especially sensitive to the plight of the rate payers.

In the long term, one can envision a scenario where electricity generation will be completely deregulated. Cogeneration and small power development is planting the seeds for such a non-regulated industry. If state regulatory agencies begin to look favorably upon such an approach in the future, both established cogeneration projects whose electricity sales contracts are expiring and new projects may be able to market their electricity product to large electricity end users as well as utilities. Utilities would then have to compete with cogenerators to retain their customers.

SUMMARY

Since the passage of PURPA, the growth of cogeneration development has been influenced by regulatory and economic forces. As a result, certain identifiable trends have been established. The reasons behind these trends provide insight into the industry's future direction.

NOMENCLATURE

N	total amount of electricity generated
Q_t	thermal energy supplied from the cogeneration cycle
Q_{fcg}, C_{fcg}	fuel energy and cost of fuel used by the cogeneration plant
Q_{fec}, C_{fec}	fuel energy and cost of fuel used by a competitive noncogeneration (utility) plant to generate electricity
Q_{fb}, C_{fb}	fuel energy and cost of fuel used to generate thermal energy by the plant other than cogeneration (i.e., industrial boilers)
η_t	overall plant thermal efficiency or fuel utilization factor (F.U.F.)
η_{et}	electrical efficiency of the cogeneration plant
η_b	steam generator thermal efficiency
η_{ec}	thermal efficiency of the condensing plant
η_{ecg}	cogeneration plant overall electrical efficiency
H.R.	cogeneration plant overall electrical heat rate

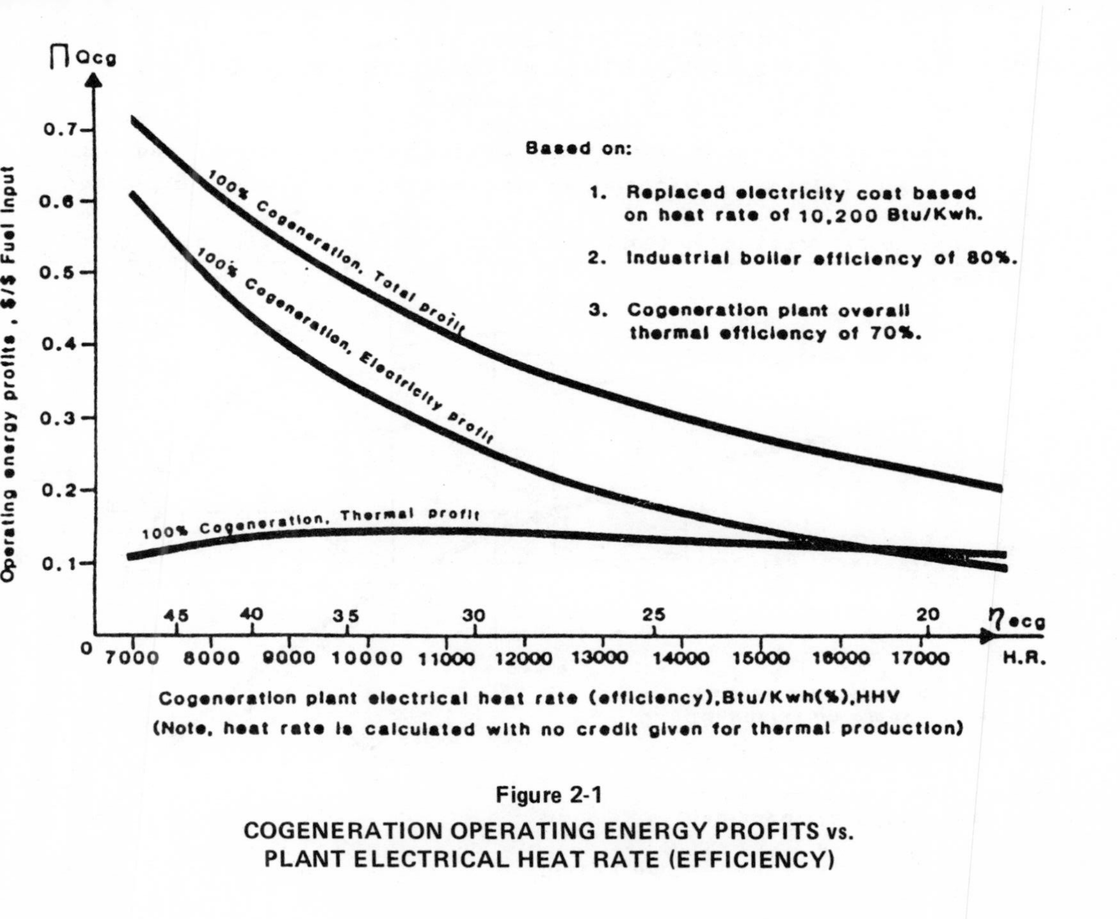

Figure 2-1
COGENERATION OPERATING ENERGY PROFITS vs.
PLANT ELECTRICAL HEAT RATE (EFFICIENCY)

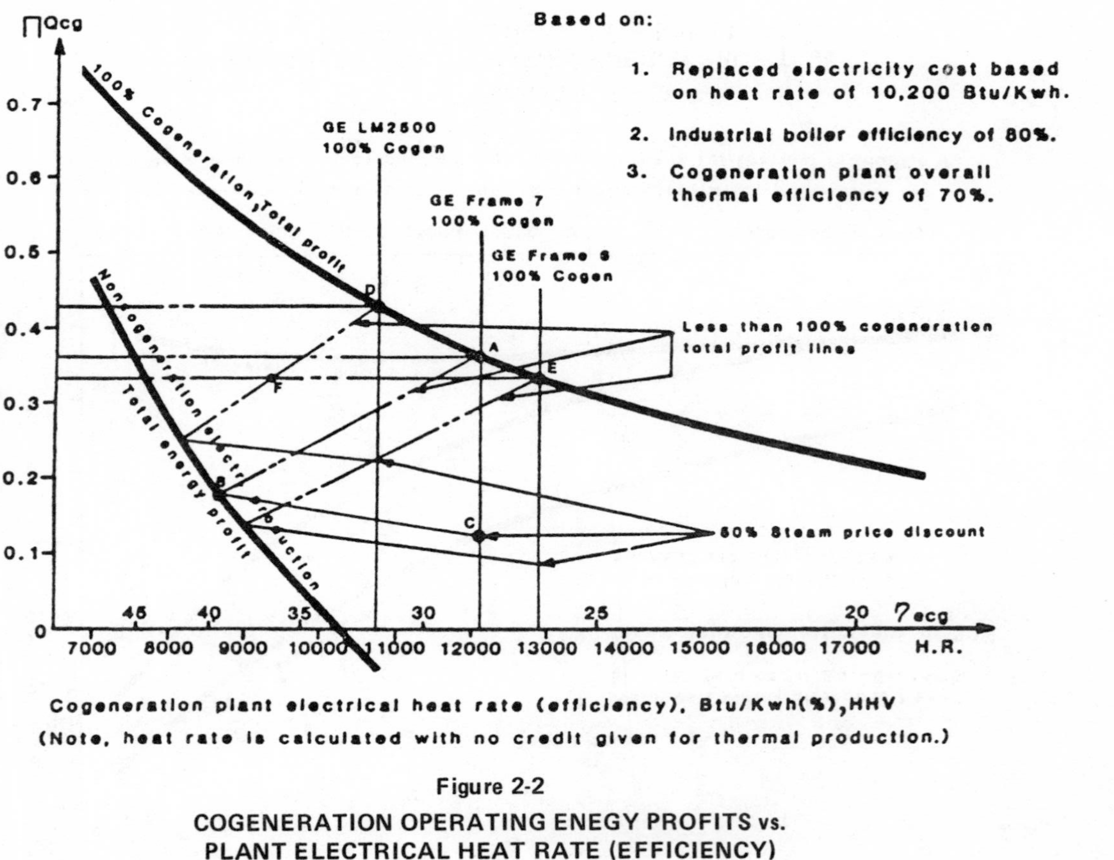

Figure 2-2

COGENERATION OPERATING ENEGY PROFITS vs.
PLANT ELECTRICAL HEAT RATE (EFFICIENCY)

PART II

ECONOMIC AND FINANCIAL CONSIDERATIONS

CHAPTER 3

An Integrated Approach to Evaluating The Technical and Commercial Options For Cogeneration Facilities In The Process Industry

D.H. Cooke and R.H. McCue

INTRODUCTION

Since November 9, 1978, the enactment date of the Public Utility Regulatory Policies Act (PURPA), there has been a strong economic and commercial impetus in the Process Industry toward cogeneration systems with high export power. The PURPA inducement is for development of plants producing electric power significantly in excess of that generated as a by-product of fuel energy released to support a given industrial process, or power consumed by the process and supporting plant systems. Additional fuel energy is released expressly to produce electric power for export, with only a small PURPA required fraction contributing to sequential production of "useful thermal energy."

The Process Industry is that sector of potential cogeneration industries which is involved with "process-stream" products, such as the chemical and petroleum refining industries. According to recent authoritative studies,[1] this sector promises to provide approximately half of the cogenerated power expected to be developed in the nation over the next ten years. Development of cogenerated power from that sector, even from retrofit of existing facilities, has been slow up to now, largely because of depressed economic conditions in the petroleum industry and legal controversy over PURPA interpretation. The opportunity for effective product diversification (into power) offered by PURPA, however, is now emerging as a stabilizing economic alternative.

Although the examples chosen for technical discussion in this chapter are typical of the chemical and petroleum refining sector, the economic evaluation approach described is applicable to any industry, including pulp and paper, food processing and steel.

A relatively low useful thermal energy output fraction is required for PURPA qualification. The thermal fraction of total sequential energy output can be as low as 5 percent under PURPA rules, although DOE rules implementing the Fuel Use Act require a minimum of 10 percent to qualify for an exemption to use oil and gas.[2] Because of this emphasis on export electric power production, the effect of the law has been to foster competition of industry with the utilities for production of consumer power, although utilities are allowed to participate (up to 50 percent ownership) in cogeneration ventures. The expressed intent of the law is to encourage *cooperative* development of cogenerated power from the utilities and industry by permitting both to benefit.

Cogeneration facilities must qualify under the PURPA rules to be eligible for "Avoided Cost" payments from the utility into whose distribution grid the power is to be fed. Additional fuel, tax and "wheeling" provisions may also be available to qualified cogenerators. Wheeling is delivery of power, through interties, to other utility grid systems serving the same markets.

While fair competition is an essential part of any major project in such a vital, complex, and necessarily coordinated field as public power, a cooperative rather than an adversary approach with the utilities will be more effective. There is no doubt that separate, independent production of heat and power is wasteful, but a return to the early days of uncoordinated production of power by many industries would be chaotic and unreliable. In order to maintain high standards for reliable power, it is necessary that the utilities remain in control as central coordinating agents for the public, and that they provide the technical criteria to be met in maintaining high system reliability and stability.

In some areas, the potential supply of cogenerated power from industry is currently greater than demand,[3] but it is reasonable to expect that as firmly committed, reliable, cogenerated industrial power is developed, it will gradually replace much of the existing uncogenerated power now in production providing the public cost is competitive. Meanwhile, the utilities are buyers in a buyer's market. Some utilities are placing the matter on a free bid basis, which, under recent Supreme Court decisions reinforcing the concept of a negotiated price,[2] the utilities are legally entitled to do. The key to a power purchase agreement between a cogenerator and a utility is to *negotiate* a price with the utility. Demonstrating that the power is *reliable* in the short and long terms, and *available*, enhances the power's value to the utility.

There are many technical, economic and financial options to be considered in structuring cogeneration projects for the most effective return on investment. This chapter describes an approach wherein technical and economic criteria are applied to size and select candidate power/process cogeneration systems, in an early stage when financial and ownership negotiations may affect, and be affected by, such sizing and selection. This integrated approach is intended to avoid the waste of financial resources which has typically resulted from downstream collapse of large cogeneration projects, for which detailed design efforts have been initiated in a volatile financial/ownership environment.

Another aspect of the integrated approach here described refers to the degree to which the power producing system is integrated with the process system yielding the commercial product. This is the *essence* of cogeneration. Yet, under the PURPA rules, the low percentage of useful thermal energy required for qualification permits power and process systems which are integrated minimally, if at all. It is believed that as cogeneration technology matures, there will be greater emphasis on integrated systems with more effective combination of capital investment toward optimized power *and* commercial product development, without the complexity of controls and duplication of components inherent in package "add-ons." In this way there can be not only conservation of energy but of financial resources. This paper will discuss various means and degrees of process/power integration, including the less integrated cycles.

From a process industry standpoint, the keys to an optimum cogeneration project lie in the ability to evaluate in detail the needs and requirements of the existing process plant (in retrofit applications), in terms of power and heat integration. Based on that evaluation, a cogeneration system may be designed that meets those needs while maintaining the necessary flexibility, reliability and operability. In-depth power *and* process technological know-how will be required to produce the most effective system.

The approach here described consists of an initial "screening" phase involving parametric analysis of technical, economic and financial options for several candidate schemes. Microcomputers are utilized during this phase for fast access/display and close coordination. Using analytical setups developed during the screening phase, a complete economic optimization of one or two finally selected systems will be accomplished in follow-on work, bringing in high-speed, high-volume mainframe computers and available multivariate optimization techniques. The cycle optimization phase is not described in this chapter, but allusion is made to certain variables for which "first guess" approximations will be assumed, to be refined later. The description in this chapter is, therefore, of the "front end" of an organized procedure making optimum use of the entire range of state-of-the-art hardware and software computing technology. The approach is not only to provide fast, flexible evaluation of the many options open in the early phases, but also to lay down a computer based "decision framework" which will allow continuing maximum access to technical, economic and financial information, and rapid recovery from inevitable changes along the successful path of an evolving cogeneration project.

THE TECHNICAL - ECONOMIC - FINANCIAL PROBLEM

The early screening phase of multifaceted cogeneration ventures is probably the most crucial. During the period when ownership, capital funding, tax ramifications, economic predictions, and power/process technical systems are all being simultaneously evaluated, there must be strong leadership and a carefully planned information/decision process. The emphasis should be on fast, but *consistent*, evaluation of promising scenarios according to rough order of magnitude. The technical systems are usually the "what" of the entire venture, and must be prominent in the evaluation from the beginning. Since the economic and financial differences between technical schemes may appear to be small, but may translate into large dollars over the life of the project, or spell success or failure, early technical insight in both the power and process disciplines is essential. A definition of the screening phase is, therefore: the selection of technical systems for further consideration and refining in later phases, and evaluation of these systems under probable ranges of economic and financial parameters.

Figure 3-1 shows a schematic of the technical-economic system to be analyzed. The technical system is defined by elements to be affected by capital investment. Cost-related inputs and revenue products are shown for a system to be retrofitted in an existing plant. The cost-related inputs and revenue products are incremental over what existed before the capital investment. Revenue includes savings from otherwise purchased power, and PURPA income from receipt of Avoided Cost.

In Figure 3-1 the fuel input is indicated as divided between process and power, although in highly integrated cogeneration systems such a division may not be obvious. Actually, the initial fuel input may be physically divided; for example, fuel supplied to an ethylene cracking furnace as distinct from fuel supplied to a linked, supporting gas turbine combustor.

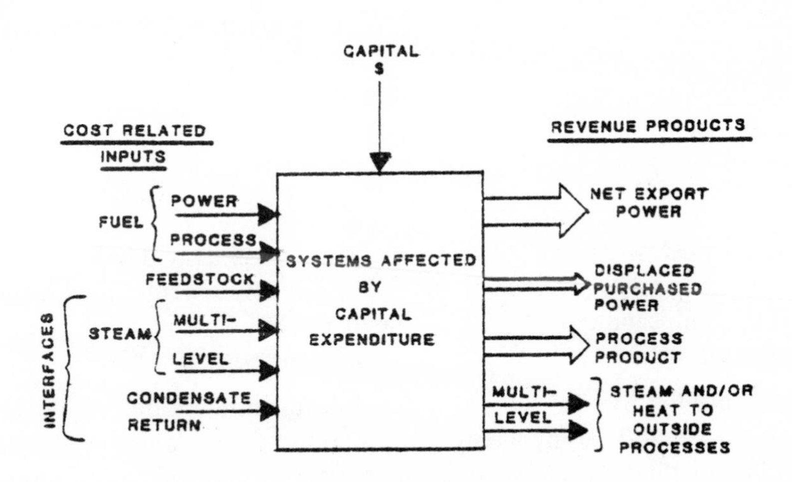

Figure 3-1
SCHEMATIC OF THE TECHNICAL-ECONOMIC SYSTEM

The Screening Analysis

Figure 3-2 shows the interrelation of the screening analysis aspects. It will be noted that the level of export power, in excess of that which is process related or consumed, is largely determined by the financial ownership evaluation.

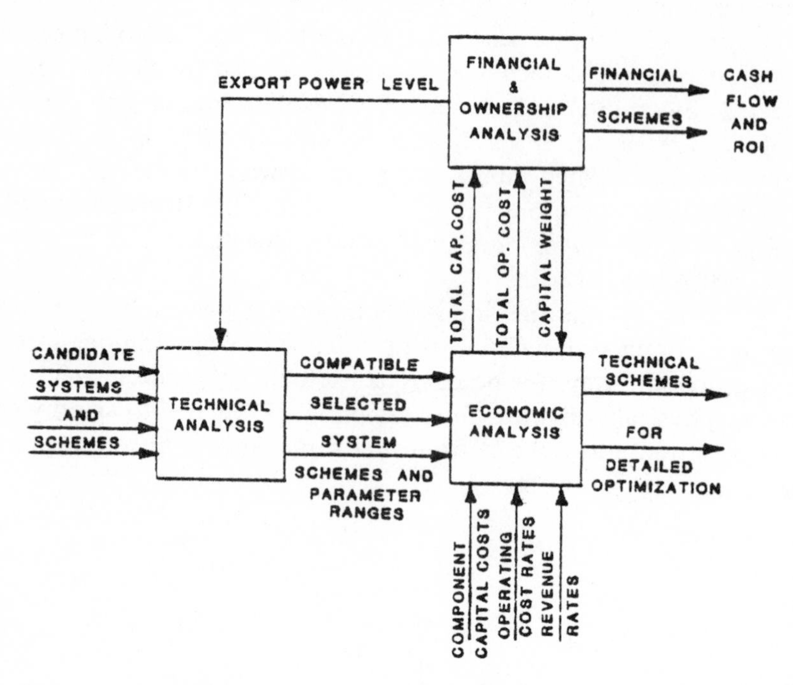

Figure 3-2
SCHEMATIC OF THE SCREENING ANALYSIS

Technical Aspects – These aspects, discussed in detail in later paragraphs, involve preselection and sizing of suitable compatible systems and candidate schemes, based on the financially attractive level of export power.

Economic Aspects – The economic parameters determine the "bottom line." This involves determination of component and total installed capital costs, system operating costs (fuel and nonfuel O&M), and product revenues affected by tradeoffs concerning critical economic and financial parameters. Financing arrangements will affect capital versus operating cost tradeoffs by influencing the weight placed on capital. The weight, or cost, of capital is determined by the investment/tax structure of the participating organizations, or of the single composite organization formed for the venture. The interest on moneys raised for the venture, as well as the required return to stockholders, must be evaluated with tax and insurance allowances to determine an *annual fixed charge* on capital investment. This is used to amortize one-time capital costs so that they may be compared properly to distributed operating costs, to determine present total (capital and operating) costs.

Although the present worth of total cost and revenue is the ultimate optimizing basis for the project, in the screening phase the "cash flow" distribution is important to establish a necessary time scale for investments and revenue returns. The establishment of cash flow is, therefore, a prerequisite for determining the cost of capital.

Financial Aspects – Financial considerations cover the economic implications of various financing and ownership options, including tax ramifications:

- Equity: This is the simplest form of financing, in which a single owner finances the entire facility. In return, he is the sole recipient of proceeds from the facility and its allowable depreciation/tax credits. Compared to other options, this type of financing usually generates a lower return on investment (ROI); however, it is a viable option when obtaining loans on favorable terms is difficult.
- Debt-Equity: This is a very attractive form of financing option as the overall high rate of return is offset by a relatively low-interest loan. The owner enjoys full benefit of the tax credits and equipment depreciation by contributing only a small portion of the facility cost. A combination of long-term and short-term loans is usually used for debt, thereby complicating the financial analysis of this option.
- Trust Fund: This is an attractive financing option because *no* equity is contributed by the owner; however, he fully enjoys the benefits of all tax credits and equipment depreciations. In this type of financing, the facilities are held in a trust and funds are borrowed as needed by the trust. Repayment of funds is a percentage of plant revenues, not to

exceed its operating income, until the debt is paid off and the trust is dissolved, whereon ownership of the facilities reverts to the owner. Thus, debt repayment is proportional to production, thereby easing debt service when the facility operates at less than design conditions. Because of the production risks involved, the financial markets generally expect a good return on investment for such projects.

- Limited Private Partnership: This is essentially an equity/debt-type arrangement in which the equity portion is shared by a number of investors. By dividing the equity into smaller denominations it is made available to a larger number of investors. Because this is promoted as a tax-shelter, it remains attractive to investors in high tax brackets who are not solely dependent on the investment. Their contribution is solicited over a period of 1 to 5 years. In return, they are allowed to take their share of tax writeoffs (i.e., tax credits, equipment depreciation and interest deductions). The financing is arranged in such a way that even though they build up equity in the venture and receive early tax writeoffs, the limited partners' net contribution remains small for the first few years. After that, such partners start receiving attractive returns on their investment. This type of financing appeals to individuals in high tax brackets, as their net contributions are moderate, while rewards are very high.

Export Power Level and Ownership – The magnitude of export power will be determined by financial considerations so as to ensure an adequate return, at an acceptable risk, attractive to investors. If the export power level is low, so that the power produced is not much above that which is process-related or consumed, the investment required for the venture can be within equity financing capability by the parent industry company, as process plant sizing is traditionally determined by this capability. For very large export power levels, the capital outlay and associated risks may be beyond the prudent capabilities of even the largest Process Industry companies, and outside investors may be desirable. Investors with enlightened interests and "embedded" assets, such as utilities with transmission line capability, may be particularly beneficial partners in these large power ventures.

Screening Phase Output

The Screening Phase Output will consist of a limited number of defined technical systems for further detailed analysis and optimization as shown in the lower part of Figure 3-2. Outputs from the financial evaluation include cash flows and ROI corresponding to the selected technical systems, as shown in the upper part of Figure 3-2.

A microcomputer program has been developed to perform economic analyses for cogeneration schemes and to evaluate various financing/ownership options. The program is flexible to incorporate variations in:

- Operating Data: Fuel costs, product revenues, operation and maintenance expenses
- Tax Laws: Tax rates, tax credits, depreciation schedules
- Finance Terms: Interest rates, closing costs

Initially, the program performs an economic analysis based on the operating data to generate "Gross Operating Income." Later, it evaluates all potential financing/ownership schemes and prepares a concise summary of each option which includes:

- Project Costs: Capital cost, construction cost, financing cost, etc.
- Financing/Ownership Details: Schedule of financing and equity contributions
- Cash Flow Details: After-tax cash flow, return on investment, payback period, etc.

The program generates financing details for any of these options when further analysis is desired.

The major required economic factors for operating a cogeneration facility are fuel costs and product revenues. In addition, their values fluctuate over the years depending upon market conditions. Therefore, the program covers a wide range of applications by including cost items like multiple types of fuel and by including revenue bearing products like steam (multiple levels), electricity, and a process product enhanced in value by alternate technical schemes. The program assumes price escalation based on industry standards. However, there is a provision to modify these factors for a particular application. Some other cost factors included in the economic analysis are operation and maintenance expenses, insurance premiums and local taxes.

TYPES OF POTENTIALLY ATTRACTIVE HEAT-POWER CYCLES

In presenting a technical overview of cycles for potential process/ power integration, we first consider two classic stand-alone types, the Furnace-Rankine and the Brayton (Gas Turbine) Combined, which have grown separately into economic compatibility with the Process Industry. A later section will show how these archetypes can be combined to considerable advantage under the new economic environment brought about by PURPA.

Fully Fired Furnace Boilers with Steam Power Only

This is the conventional Rankine cycle, "power-boiler" system used widely and improved up to 1975 by many industries. Pre-PURPA economics established this system as the most frequent cycle-of-choice because of lower operating (fuel) costs in a period when the major electric economic issue was whether to self-generate or buy power to be consumed on site, and the thought of exporting power was out of the question. With economic throttle steam conditions up to 1,450 psig, 900°F, the most economical systems of 1975 vintage were capable of providing enough power for the plant's own use, while supplying, as a primary objective, large quantities of steam to process. Figure 3-3 shows a typical cycle of this era.[4]

The "fully fired" characteristic of the power boiler refers to the use of minimum excess air for complete combustion, i.e., about 10 percent above stoichiometric air. This usually produces furnace temperatures above 2500°F, requiring refractory construction and heavy steel support. Although the combustion is complete and efficient, steam turbines are not able to utilize temperatures this high, and gas turbine cycles with higher working fluid temperature capability are tending to replace the stand-alone furnace steam boiler in post-PURPA designs in which power is the primary object. However, the furnace boiler is still a necessary and economical design in certain processes and applications, as described below.

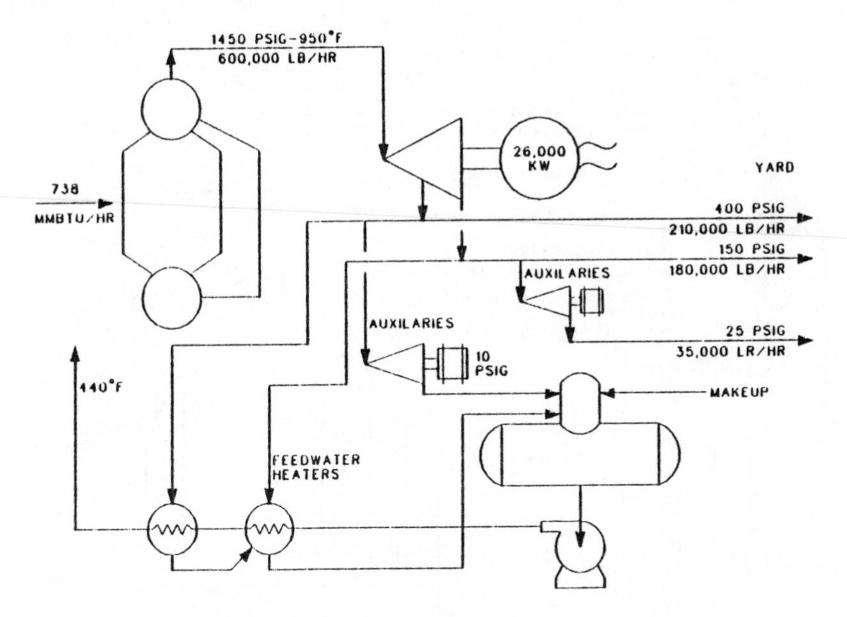

Figure 3-3
FURNACE-RANKINE STEAM POWER CYCLE

Process Furnace – In cracking furnaces typical of olefins plants, and fired heaters typical of refinery plants, the high temperatures required for the process or thermochemical reaction frequently demand a refractory or radiant furnace environment. Figure 3-4 shows a typical olefins cracking furnace and related heat recovery systems. After accomplishing the cracking reaction at a carefully controlled high temperature (up to 1600°F) in the feedstock tubes in the radiant "firebox" of the furnace, heat recovery takes place both in the convective section of the furnace and in external process stream coolers. The rapid cooling of the process stream by boiling feedwater in specially designed external coolers, as shown in Figure 3-4, is critical to stop the cracking reaction at precisely the design "residence" time. The residence time of the process fluid at temperature and the reaction temperature of the process fluid in the radiant tubes are two of the most critical parameters for process design and control.

Steam generation is done primarily in the process stream coolers in a cracking furnace because of the high cooling rate required. The

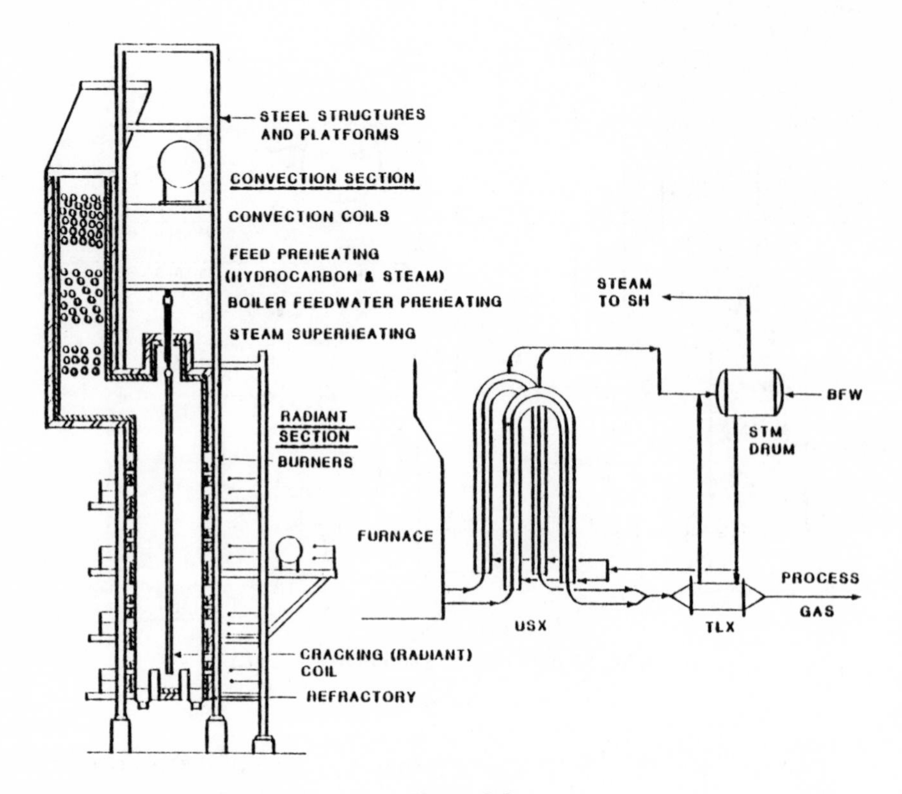

Figure 3-4
CRACKING FURNACE AND PROCESS STREAM COOLING SYSTEM

process stream cooling system shown in Figure 3-4 is a highly efficient and operationally trouble-free system known as USX-TLX[5] (for Ultra-Selective and Transfer Line Exchangers). It is applicable for rapid, efficient process stream cooling in many cracking processes and is capable of generating high-pressure steam. Heat recovery in the convective pass of the furnace is accomplished by combustion air, feedstock and boiler feedwater preheating and steam superheating.

In order to provide even temperatures in the reacting process fluid, with low peak-to-average temperature ratios, the tubes in a cracking furnace are placed in a single row configuration in the flame-center of the long, narrow, insulated firebox. In a fired-heater firebox (Figure 4-5), process temperatures are less critical, and wall tubes are utilized. The fired-heater is a "workhorse" component in many process industries. A typical application example is crude

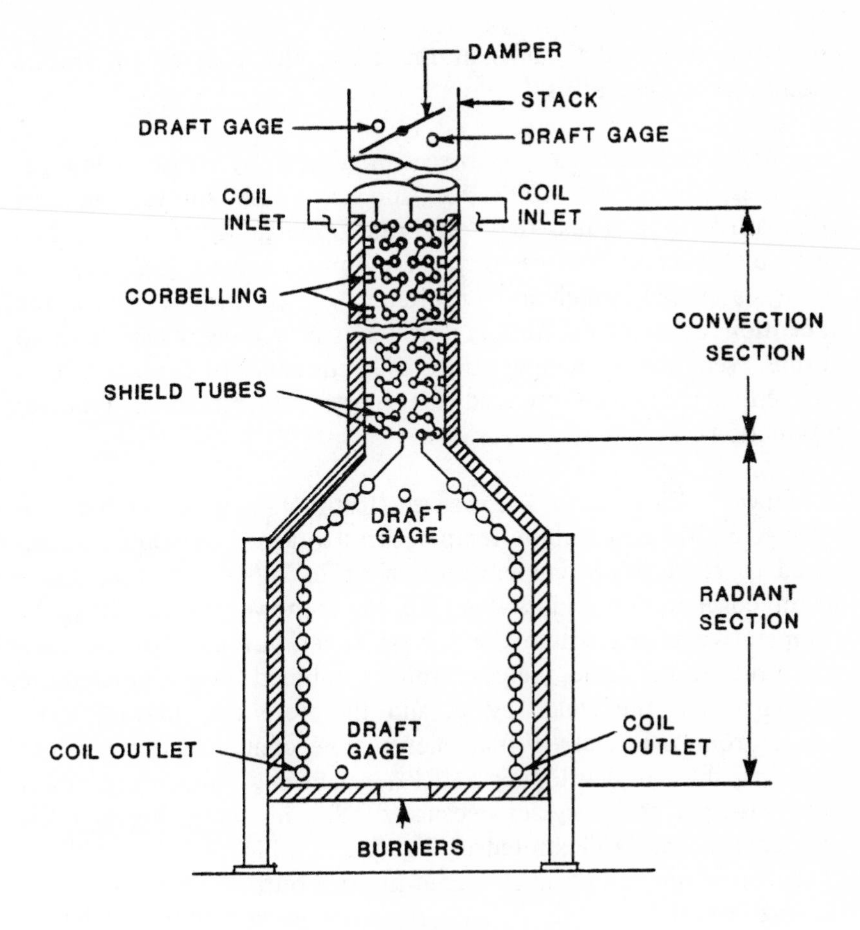

Figure 3-5
FIRED-HEATER FURNACE

heating in a refinery in preparation for fractionation. Heat recovery in the fired-heater takes place in the convective pass only, and usually includes steam generation and feedstock preheat.

In cracking or fired-heater furnaces, the generation of steam, which can be at high pressures and temperatures (up to 1450 psig, 950°F), is a necessary and efficient heat recovery measure, providing energy for sequential utilization as heat, or for turbine power. In contrast to the power boiler, the steam drum is usually external to the furnace, with natural circulation downcomers and steam piping conecting it to tube banks in the convective pass or to process cooling steam generators.

Fuel Considerations – Furnaces in general, by virtue of low gas velocities, which cause no erosion, and ease of ash removal, can burn "difficult" fuels. Thus, they offer the opportunity to reduce fuel costs by utilizing coal or low-cost by-product fuels (e.g., coke or pyrolysis pitch), which are unsuitable for gas turbines. With the exception of cracking furnaces, which still require clean fuels to ensure even firebox temperatures, the economics of furnace boilers frequently emphasize reduced fuel costs rather than high power production.

Power-to-Heat-Ratio – Because stress-metallurgy economic constraints limit steam turbine temperatures to less than 1000°F, compared to gas turbine temperatures of up to 2000°F, furnace boiler steam cogeneration plants tend to lower power-to-heat-ratios. In comparative plants where the lowest steam turbine exhaust temperatures are the same, the gas turbine combined cycle is much more efficient than the steam cycle, and therefore can produce more power from the released fuel energy. The combined cycle is also more capable of developing PURPA qualifying condensing power than gas-fired steam cycles because of the inherently higher cycle efficiencies. This is illustrated in Figure 3-6, which shows that steam cycles must operate at much higher thermal output (heat-to-process) in order to qualify under PURPA rules for gas and oil fuels.[6] This is one of the main reasons that gas-turbine combined cycles are more suitable for cogeneration systems where the economic emphasis is on power, and gas fuel must be utilized.

PURPA Rules for Solid Fuels – Although steam plants have lower efficiencies, condensing power can be utilized freely in solid fuel plants, even at low thermal output, because there are no PURPA efficiency standards where gas and oil fuels are not involved. Heat rejection by condensing reduces thermal efficiency to values approaching 35 percent, which is typical of the best noncogenerated industrial power plants. The condensing capacity available with solid-fuel-fired Rankine cycle steam plants provides flexibility for variable process steam utilization, allowing large Avoided Cost

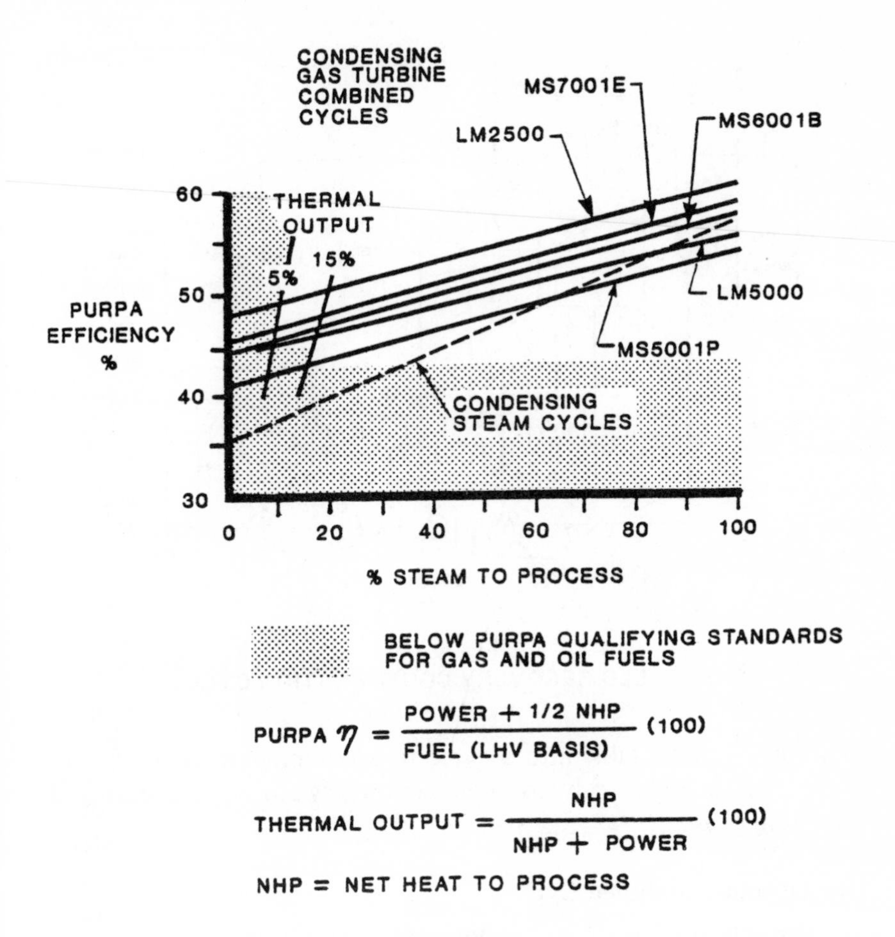

$$\text{PURPA } \eta = \frac{\text{POWER} + 1/2 \text{ NHP}}{\text{FUEL (LHV BASIS)}} (100)$$

$$\text{THERMAL OUTPUT} = \frac{\text{NHP}}{\text{NHP} + \text{POWER}} (100)$$

NHP = NET HEAT TO PROCESS

Figure 3-6
PURPA EFFICIENCY AT VARIOUS PROCESS STEAM DEMANDS

revenues to flow in during periods of process turndown operation (low thermal energy output). An example of this is shown in Figure 3-7. The cycle shown there[7] was initially designed to produce a high flow of low quality steam for oil field recovery enhancement. During later years of the plant life when the oil field is depleted, the cycle would be converted to almost full utility power production and ownership (still retaining the PURPA qualifying 5 percent useful thermal energy), with additional condensing units added to utilize the full boiler steam production capacity. This is an example of how

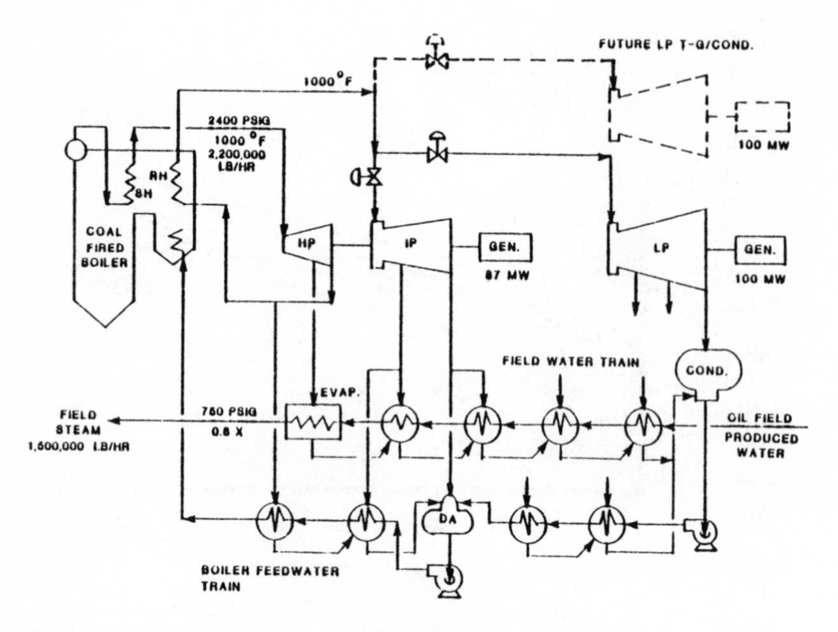

Figure 3-7
OIL FIELD RECOVERY COGENERATION CYCLE

flexibility can be built into a cycle to accommodate changes necessitated by financial and economic conditions anticipated during the screening phase.

Gas Turbine Combined Cycle

The gas turbine (GT), or Brayton, cycle can be used in many combinations with other cycles, including the furnace-Rankine cycle described in the preceding section. Among these combinations are cycles utilized in large-scale coal gasification/liquefaction plants which have received less attention recently due to the fall in oil prices. In all such combinations the ability of state-of-the-art gas turbines to operate reliably at inlet temperatures up to 2000°F, with well-developed blade cooling techniques, is utilized to achieve combined cycle efficiencies much higher than that of the best steam turbine cycles. This can be seen in Figure 3-6, where a high-efficiency, aircraft derivative (LM2500) GT-based combined cycle shows an efficiency of 48 percent with no process heat recovery. This compares with about 35 percent for the best industrial boiler steam cycles, even with regenerative feedwater heating, primarily because the steam turbine is limited in inlet temperature to 1000°F or less.

The steam cycle is limited to 1000°F not by the turbine bladepath, since the same super-alloy materials, structure and blade cooling techniques used in gas turbine blading and rotors could be used in high-temperature steam turbines. The 1000°F limit is due to the high *pressures* and thick-walled turbine casings, boiler steamdrums and piping necessary for steam at high temperatures. Due to a change in design-strength metallurgy from ferritic to austenitic at slightly above 1000°F, the high cost of field and factory fabrication of massive austenitic pressure vessels, as well as large quantities of similarly massive piping, valves and fittings has long made high-temperature steam plants uneconomical. The gas turbine Brayton cycle enjoys the capital cost benefit of low turbine casing and duct pressures, with the highest pressure usually less than 300 psia and localized in the vicinity of the combustor.

The temperature-efficiency advantage of the combined cycle gas turbine translates directly into increased power which can be seen dramatically in the temperature-entropy diagram shown in Figure 3-8.[8] Since the crosshatched areas in the diagram are equivalent to the thermal energy converted to power, it can be seen that the Brayton cycle, superposed on the Rankine cycle, converts fuel energy to power at temperatures unattainable in the steam cycle.

The GT/HRSG Cycle – An example of the gas turbine-heat recovery steam generator (HRSG), the most popular stand-alone cycle in many industries including Process, recently, is shown in Figure 3-9.[8] This cycle has been called the "PURPA Machine" because of its fast-build characteristics, relatively low capital cost requirements, and ability to burn extra fuel for *large* electric power production. The cycle in Figure 3-9 typically has a relatively low temperature (~1000°F) gas turbine exhaust, which permits the use of the light construction, compact-heat-exchanger type, low-cost, steam generator known as the HRSG (Figure 3-10).[10] The gas turbine exhaust can be "supplementary fired" to raise the HRSG inlet gas temperature as high as 1600°F, still below that requiring use of refractory, cooled-wall construction. This further improves cycle efficiency and permits great flexibility in process steam production control and/or greater steam turbine power.

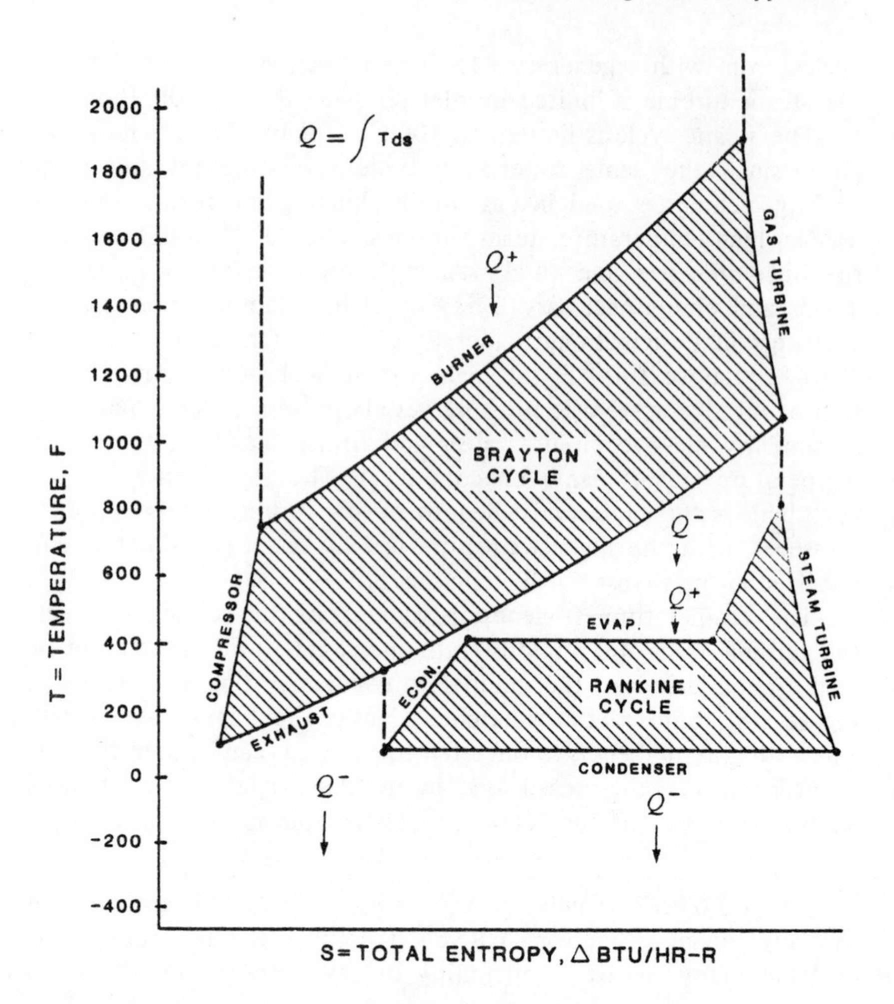

Figure 3-8
TEMPERATURE-ENTROPY DIAGRAM GT-HRSG COMBINED CYCLE

Fuel Considerations – State-of-the-art gas turbines are currently limited to high-cost, critical, restricted fuels (gas and oil). Provided that a fuel saving can be demonstrated, the Fuel Use Act permits qualifying PURPA cogeneration facilities to use gas as a fuel. However, this is forbidden to utilities developing new base loaded and peaking capacity without cogenerating.

It should be pointed out here that the limitation of the gas turbine to delivered gas and oil fuels may not be long in passing.

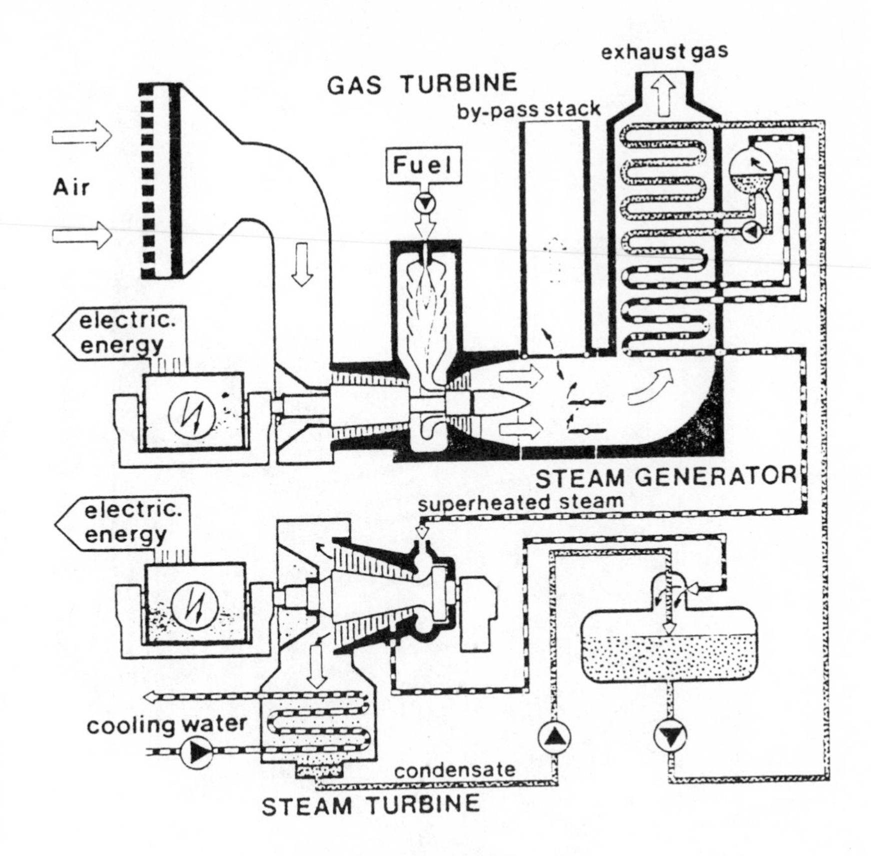

Figure 3-9
GAS TURBINE-HRSG CYCLE

Although direct firing of coal in GT combustion has not been promising, recent developments in fuel system integration of small-scale coal gasification processes (e.g., British-Gas/Lurgi) may facilitate a breakthrough.[8]

Oil- or gas-fired gas turbines provide a proven, reliable technology with simple, low-capital-cost, quick-build characteristics. The gas turbine also provides a wide range of process integration possibilities beyond the use of HRSGs, which is discussed in a later section.

PROCESS INTEGRATION SCHEMES WITH GAS TURBINES

The Furnace Rankine and gas turbine combined cycles described in the preceding section can be integrated in a variety of ways in a

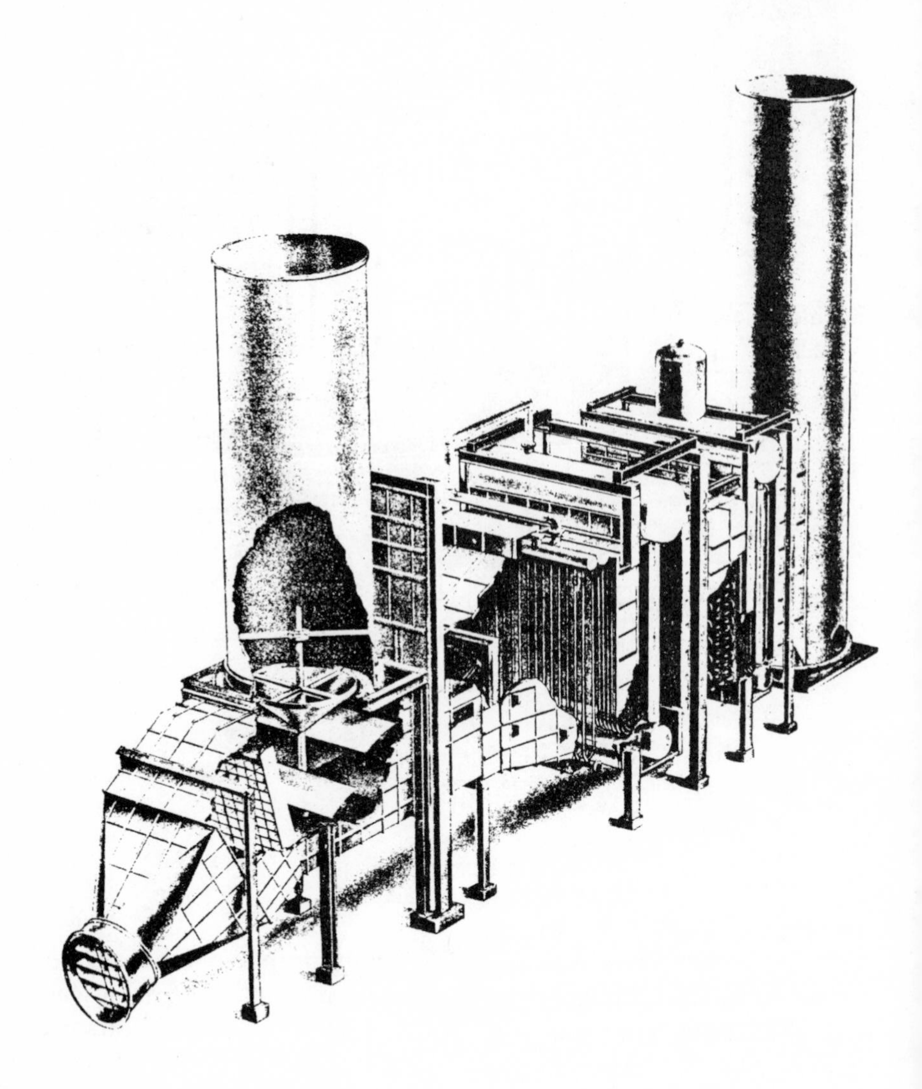

Figure 3-10
HEAT RECOVERY STEAM GENERATOR

Continued ⟶

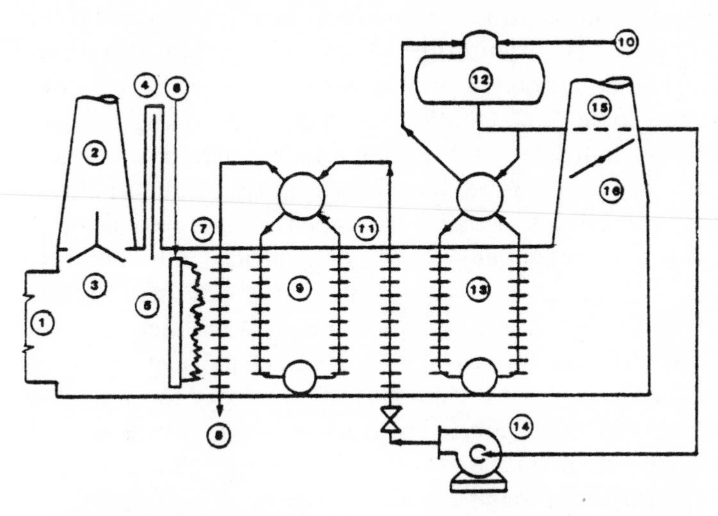

1. GAS TURBINE EXHAUST
2. DUMP STACK
3. DUMP VALVE
4. ISOLATION GATE
5. BURNER GRID
6. FUEL TO BOILER
7. SUPERHEATER
8. STEAM OUTLET
9. HIGH PRESSURE BOILER
10. CONDENSATE RETURN
11. ECONOMIZER
12. DEARATOR
13. L-P BOILER
14. BOILER FEED PUMP
15. MAIN STACK
16. VALVE TO INCREASE DRAFT LOSS TO ASSIST DUMPING

Figure 3-10 – Continued

process plant where furnaces are required for high-temperature reactions or processes. The gas turbine provides the most effective means of producing power for export, and the oxygen-rich turbine exhaust gas can be utilized as preheated combustion air for the furnace, and/or separately cooled in an HRSG to generate high-pressure steam for process use or steam turbine power. The objective in combining these systems is to reduce the total fuel required for product and power production in order to maximize the return on investment. Most schemes shown in this section are from Reference 11.

Basic Process Integration Schemes

The schemes which follow describe modules consisting of a single gas turbine in combination with a furnace and an HRSG. This is the "basic" combination, although in large processes there may be multiple furnaces and multiple GT-HRSG units. The gas turbine and HRSG should always be linked as a set, because experience has shown this to be an economic and practical operating arrangement.[10] Manifolding multiple gas turbines into a single downstream HRSG poses the danger of hot exhaust from an operating GT flowing back through another GT which has tripped off, as well as problems in restarting a shutdown engine. A safe solution with dampers is complicated and costly compared with one-on-one modularization for the GT-HRSG, which has become a standard practice in the industry.

When multiple furnaces or divided GT exhausts are required, it will be necessary to provide shut-off dampers to ensure operating reliability and flexibility. Furnaces in large processes are often modularized in manifolded groups to permit ease of process and plant control, and progressive maintenance.

In all of the basic combinations shown in this section, the gas turbine can be shut down without adversely affecting the process, although backup systems (e.g., standby fans for combustion air) may be needed to ensure this in some instances. The schemes also permit sizing of the gas turbine without limitation by the process thermal duty. Any size gas turbine can be utilized up to the maximum available in the market, or multiple GT-HRSG sets, according to the level of export power desired, and governed by PURPA efficiency requirements (Figure 3-6) when the total fuel heat-release exceeds plant capability to utilize rejected heat.

Minimum Integration System – The system shown in Figure 3-11 is the least integrated, typical of a package GT-HRSG added outside the battery limit (OSBL), to an existing process plant consisting of one or more furnaces. Thermal integration of power and process systems may involve only minimal integration, such as partial feedwater heating using extraction steam, to provide with the process steam, the minimum Fuel Use Act required 10 percent useful thermal output of the total energy produced.

Integrated Steam Systems – Figure 3-12 shows the process and power related steam systems integrated, utilizing the HRSG for all process feedwater heating and superheating. The location of the GT-HRSG system inside the battery limit (ISBL) makes this integration economical by avoiding long steam piping runs. Additional

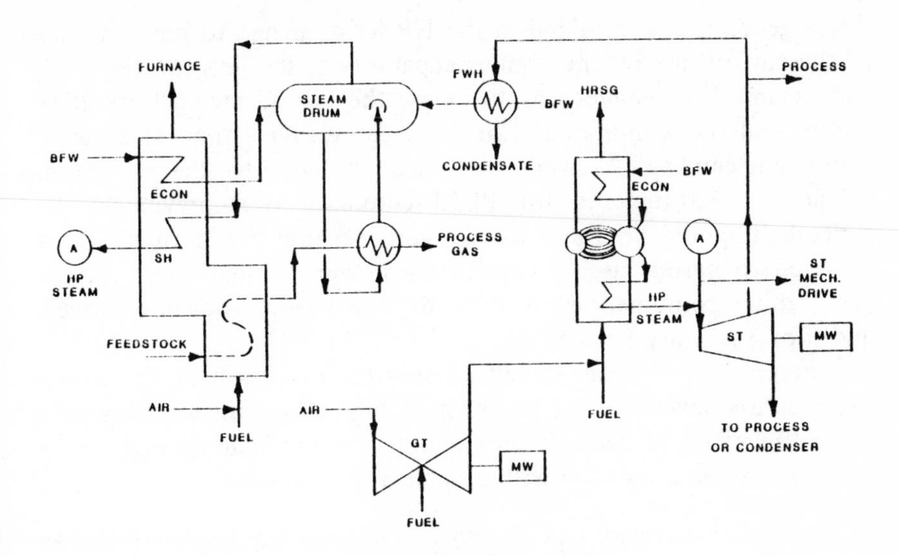

Figure 3-11
MINIMUM INTEGRATION SYSTEM – OSBL

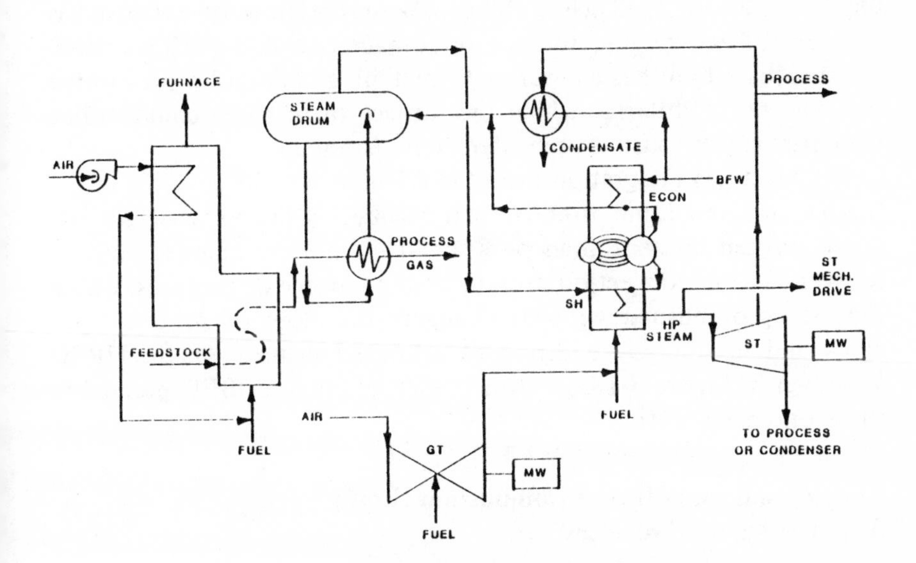

Figure 3-12
STEAM SYSTEMS INTEGRATED – ISBL

steam generation is retained in the HRSG so as not to limit the size of the gas turbine by the cooling capability of the process feedwater and steam. For smaller gas turbines, the HRSG steam generating section can be eliminated. For large gas turbines the extra steam generated can be "power condensed" subject to the minimum efficiency requirements for PURPA qualification previously described. It should be noted from Figure 3-6 that the thermal output required to permit use of condensing power is much lower for the gas turbine combined cycle than for the lower efficiency process and supplementary fired steam.

Integration of steam systems makes the process furnace convection section available for combustion air preheating resulting in a more fuel-efficient process combustion system. Again, one or more furnaces can be integrated with a single gas-turbine set.

Integrated Exhaust Gas Systems – Preheated combustion air can be provided to the process furnace or furnaces without fans by integration of the oxygen rich gas turbine exhaust as shown in Figure 3-13. The system shown is for a large gas turbine, where exhaust gas exceeding the capacity of the furnace or furnaces is utilized in the HRSG. For small gas turbines, the entire flow of exhaust gas may be utilized by the furnace or furnaces, with the HRSG eliminated. This establishes an arbitrary "matching" minimum gas turbine size for the module for which the furnace or furnaces could utilize the entire GT exhaust as preheated combustion air.

Exhaust gas integration from the GT increases the gas flow in the furnace for the same firebox heat release, as the exhaust gas has lower oxygen content than pure air. This requires augmented heat recovery in the convection section of the furnace or furnaces, which can be accomplished without changing the flow of process feedstock and product by addition of saturated steam from the HRSG as shown in Figure 3-13, or by a higher quantity of BFW preheat in the economizer section.

Furnace and Fired-Heater Combustion Group Sizes for Typical Processes

As an aid in matching minimum gas turbine output to the foregoing systems, Table 3-1 shows ranges of world class furnace combustion group sizes for cracking furnaces and fired-heaters of various

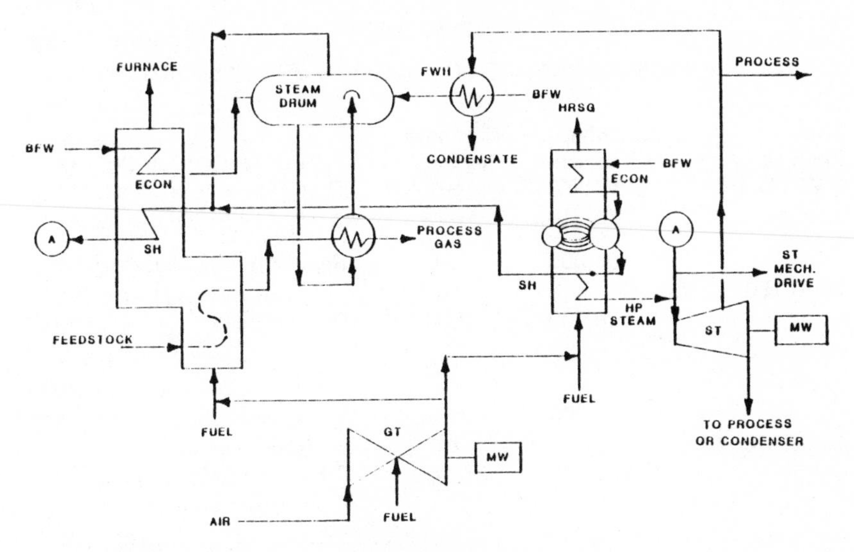

Figure 3-13
GAS SYSTEM INTEGRATED – ISBL

processes. "World class" capacities are generally recognized in the Process Industry as the economical plant sizes for various processes according to plant control, batch quantities and maintenance requirements. These are subject to change with evolving technology, but Table 3-1 gives an idea of the relative furnace group sizes, and matching minimum gas turbine outputs.

Table 3-1 shows that the minimum gas turbine sizes to match the combustion air requirements of the processes are quite small. Current industry practice has been to provide steam requirements of the entire plant with GT-HRSG units. Gas turbine sizes determined in this manner are up to 4 times the minimum GT sizes shown in Table 3-1, without the use of condensing steam power. Figure 3-14 shows a large-scale olefins cycle currently under consideration, which utilizes two 70 MW GT-HRSG trains.

POWER CYCLE/COMPONENT OPTIONS

Integrated power and process systems have been considered on an overall cycle basis. Some of the more detailed power component and cycle subsystem options should now be discussed in order to indicate the best available systems.

Table 3-1
PROCESS FURNACE COMBUSTION GROUP SIZING

Total Fired Duty, MMBTU/HR	Matching GT Exhaust Flow, LB/HR	Minimum Gas Turbine Output, kW	Type of World Scale Process That Typically Fits Duty Range
75 to 150	60,000 to 120,000	1,000 to 3,000	Reduced crude tower feed heater Catalytic cracking preheater Delayed coker
150 to 375	120,000 to 320,000	3,000 to 10,000	Atmos, crude tower feed heater Visbreaker Hydrocracker Ethylene dichloride cracker Individual olefins furnace
375 to 1,000+	310,000 to 1,000,000+	10,000 to 35,000+	Styrene preheater Total olefins cracking plant Ammonia plant Hydrogen steam reformer

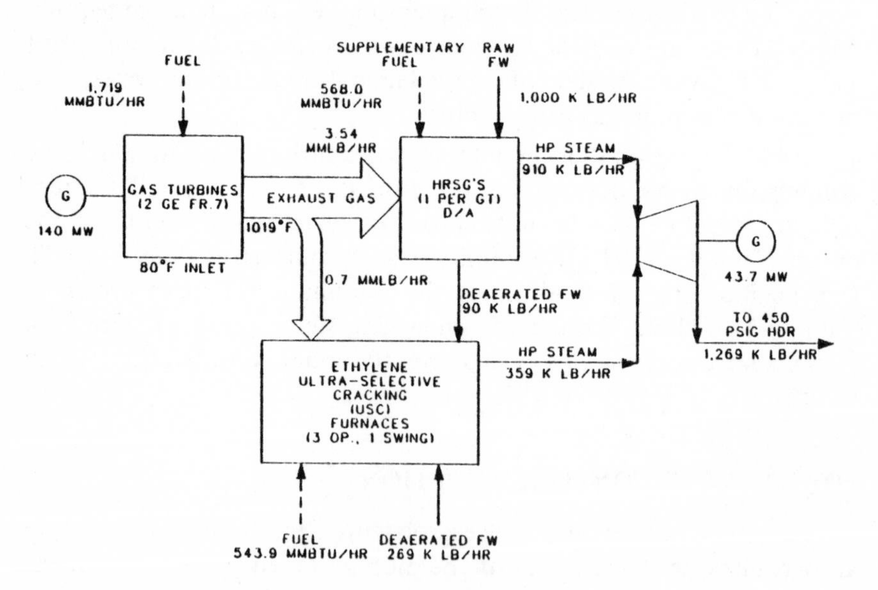

Figure 3-14
COGENERATION AND ETHYLENE INTEGRATION

Gas Turbines

Gas turbine size is the most important step in determining a financially attractive level of export power. The gas turbine usually represents the largest single investment required, and the level of export power is largely determined by the GT selection in plants of this type. Since gas turbines are limited in size by market availability, a review of commercially available designs is in order. Table 3-2 shows the technical characteristics of gas turbine gensets now on the market worldwide which are in the required power range, with 60 hz frequency, suitable for the larger cogeneration plants discussed. Data in Table 3-2 are from Reference 12.

Several manufacturers offer units above 70 MW. The GE Frame 7 at 76.5 MW has received attention lately for Process Industry plants similar to the ones discussed. The largest gas turbine unit considered technically feasible is around 100 MW in generator output. Only one manufacturer has offered a well proven 60 hz unit in this range, the Westinghouse W-501D. A new prototype is now on the market from Kraftwerk Union (Siemens), the V84.2, which is slightly smaller at 92 MW.

The larger GT units are all industrial-type machines with heavy rotor construction and longer mean-time-between-overhaul (MTBO) than the more efficient aircraft derivative designs. Industrial GT designs are also available in the smaller sizes, as with the GE Frame 3 rated at 10.5 MW.

Heat Recovery Steam Generators (HRSG)

Well proven light construction HRSG designs are available from many competent firms to accommodate the largest exhaust flows from gas turbines shown in Table 3-2. Because the gas temperatures are relatively low (800°F to 1600°F), finned tubes can be utilized to increase surface heat transfer effectiveness, reducing the required surface area and cost.

Single versus Multi-Drum – The "pinch point" in an HRSG limits the amount of steam which can be generated at a given pressure level as shown in Figure 3-15. It is thermodynamically more efficient to generate steam at several pressure (and temperature) levels, as shown in the figure, to more closely parallel the decreasing temperature gradient in the gas stream in "stair-step" fashion. This increases the

Table 3-2

GAS TURBINE GENSET UNIT PERFORMANCE DATA (ISO CONDITIONS) BASE LOAD, 60 HERTZ, 19000 kW and ABOVE

Manufacturer	Model		Gross Power kW	Heat Rate Btu/kWh	Exhaust Flow lb/hr	Exhaust Temp. F
Westinghouse CTS	W-501D		102965	10220	2988000	950
Kraftwerk Union	V84.2		92000	10330	2728800	954
General Elec GTD	MS7001E		78700	10590	2192400	995
BBC Brown Boveri	Type 11		72500	10675	2268000	968
BBC Brown Boveri	Type 8		46400	10660	1404000	977
Westinghouse CTS	W251B		42075	11337	1346400	995
Fiat TTG	TG 20		39410	11130	1278000	975
General Elec GTD	MS6001B		38100	10765	1083600	1006
BBC Brown Boveri	Type 9		35400	11810	1292400	950
Ingersoll-Rand	GT-71	*	33918	9045	1036800	785
Ishikawajima-Har	IM5000	*	33200	9080	986400	788
GEC Gas Turbines	ELM150/2	*	33010	9158	990000	797
UTC Power Sys	FT4C-3F	*	31350	10648	1083600	849
Rolls-Royce IMD	SK30	*	28100	11150	856800	1030
General Elec GTD	MS5001P		25890	12110	961200	916
Curtis-Wright	Mod Pid 25	*	24900	11850	849600	968
GEC Gas Turbines	EO-1C	*	24040	10936	374400	910
Dresser-Clark	DJ-290R	*	21998	9776	705600	832
Sulzer Brothers	Type 10		21300	10160	583200	945
Ingersoll-Rand	GT61	*	20988	9594	529200	964
General Elec GTD	LM2500-30	*	20507	9246	525600	920
Dresser-Clark	DJ-270G	*	20432	9414	518400	930
Ishikawajima-Har	IM2500	*	19200	9610	511200	930

*Aircraft Derivative

average temperature (and energy availability) of the generated steam. Multiple steam drums are used to accomplish this in HRSG's, and up to three drums have been utilized to economic advantage. Many systems utilize the low-pressure steam to provide deaeration for feedwater as illustrated in the two-drum system in Figure 3-10. Multiple drum systems can be utilized with progressively lower steam turbine injection to improve cycle efficiency.

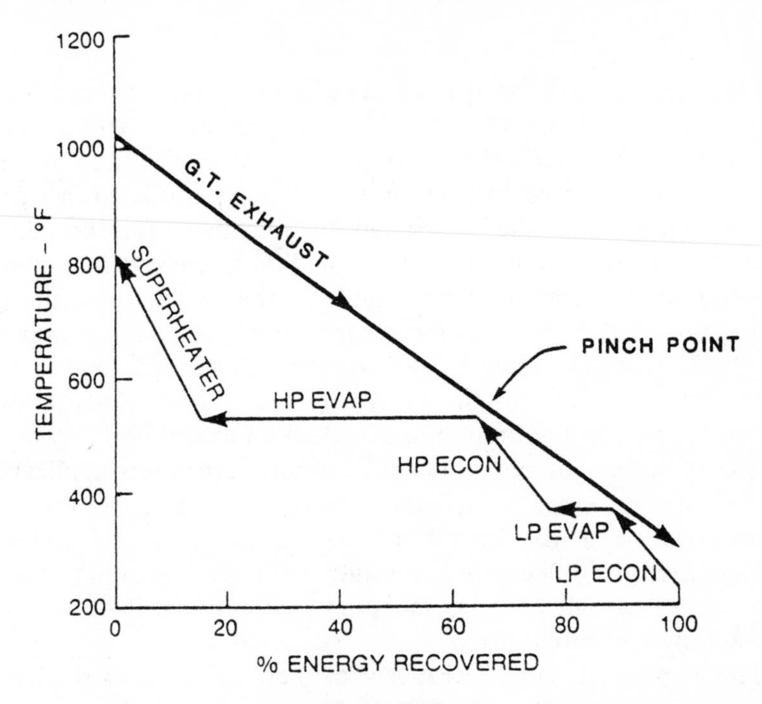

Figure 3-15
MULTI-DRUM HRSG TEMPERATURE GRADIENTS

Supplementary Firing – In order to limit turbine blade operating temperatures, gas turbines use large amounts of excess air in their combustors. Typically, gas turbine exhausts contain about 15 percent oxygen by weight, [10] compared to 23 percent in air and 2 percent in flue gas from fully fired furnaces. This oxygen is adequate to support additional or supplementary combustion. Supplementary firing can be utilized to produce additional steam from the HRSG for process and/or power use. With the large "condensing margin" available with gas turbine combined cycles, as indicated in Figure 3-6, much of this steam can be used to produce condensing power within PURPA rules.

HRSG Inlet Temperature Control – In addition to providing additional steam and power, supplementary firing provides more flexibility in process steam control. With and without supplementary firing, the gas turbine Inlet Guide Vanes (IGVs) can also be used to control HRSG inlet gas temperature.

CONCLUSION

This chapter has outlined a number of possible financial and technical arrangements for developing high export power cogeneration systems in the process and refinery fields. Many applications in these fields are characterized by the continued requirement for gas-fired furnaces because of the controlled high temperatures needed in commercial processes. Because of its ability to convert fuel energy to power at temperatures unattainable in the steam cycle, the gas turbine provides the most effective means of producing power for export in gas-fired systems. The oxygen-rich exhaust makes it a natural topping cycle for the process furnaces where clean burning can be accomplished with little or no effect on the process.

The optimum combination will be different for every application and industry in every utility area. Discovering the most effective export power level will depend on a careful screening analysis based on a *cooperatively negotiated* position with the controlling utility company. Financial, process, and power expertise will be required to effect these negotiations.

The impact of the PURPA law on new utility construction is evident to anyone who can read a newspaper. Yet, the potential for coordinated, cogenerated industrial power growth is real, particularly in the southwest region, where heavy industry remains a viable economic force.

ACKNOWLEDGEMENTS

The assistance and support of Fred Hassid, Bharat Gandhi, Moses Koeroghlian, Jay Ross, and Krishna Merchant of the Stone & Webster Power and Process Engineering staffs is gratefully acknowledged.

REFERENCES

[1] P. Bos, "The Potential for Industrial Cogeneration Development to Defer New Utility Electric Power Capacity over the Next Decade," Resource Planning Associates, June 1983 (DOE Publication No. RA-83-0107).

[2] M.M. Schorr (General Electric Co.), "The Impact of Legislation on Cogeneration System Development," ASME Paper No. 85-Pet-5, Energy Sources and Technology Conference, Dallas, February 1985.

[3]J.L. Blau (Houston Lighting & Power Co.), "Cogeneration in the Real World," Panel Discussion at International Gas Turbine Conference (IGTC), Houston, March 1985 (Unpublished).

[4]B.J. Mangione and J.C. Petkovsek (Mobil R&D Corp.), "Economics Favor Industrial Power Generation with Steam from High Pressure Boilers," ASME Paper No. 75-IPWR-5, Industrial Power Conference, May 1975.

[5]L.E. Chambers and W.S. Potter (Stone & Webster Engineering Corporation), "Design of Ethylene Furnaces," *Hydrocarbon Processing,* January-August 1974.

[6]J.M. Kovacik, "Industrial Gas Turbine Cogeneration Application Considerations," General Electric Co. Publication No. GER-3430, 1983, Figure 11.

[7]Application for Certification,Belridge Field Cogeneration Project (to California Energy Commission), Bechtel Power Corporation for Shell California Production, Inc., December 1982.

[8]R.H. Kehlhofer (BBC, Ltd.) "300 MW Combined Cycle Plant with Integral Coal Gasification," ASME Paper No. 84-JPGC-GT-19, October 1984.

[9]*Cogeneration and Small Power Monthly,* December 1984, page 2.

[10]R.W. Foster-Pegg (Struthers Energy Systems, Inc.), "Gas Turbine Heat Recovery Boiler Thermodynamics, Economics and Evaluation," ASME Paper 69-GT-116, March 1969.

[11]W.V.L. Campagne (Stone & Webster Engineering Corporation), "Gas Turbine Selection for the Chemical Industry," A.I.Ch.E. Paper No. TP84-76, Summer National Meeting, Philadelphia, PA, August 1984.

[12]Editors, *Cogeneration* and *Gas Turbine World* magazines, *Industrial & Marine Gas Turbine Engines of the World 1985-86,* January 1985.

CHAPTER 4
Cogeneration Planning

Martin A. Mozzo, Jr., P.E.

INTRODUCTION

The term "cogeneration" is rapidly becoming a new buzzword in the energy manager's vocabulary. Many managers are actively pursuing cogeneration as part of their energy programs. However, it has been my experience that some individuals do not fully comprehend what cogeneration really is and its ramifications to them. Instead, they have a project which is not completely thought out and thus subject to failure. This chapter discusses why American-Standard is considering cogeneration at some of its key plants, and some of the key considerations which must be evaluated in such projects.

WHY COGENERATE?

Cogeneration is defined as the sequential generation of electrical and thermal power using one energy source. Cogeneration is not a new technology, but rather one that was prevalent at the turn of the century. According to a recent Office of Technology Assessment (OTA) report, over 59 percent of total U.S. electric generating capacity in 1900 was at industrial sites. This was primarily because electric service was unavailable, unreliable, and costly. Introduction of reliable and reasonable electrical service in the mid-1900s, due to economies of scale for central power plants, led to the disappearance of industrial cogenerators.[7] It has only been in recent years that cogeneration is widely being considered again as a viable technology.

There are four (4) key factors which are renewing interest in cogeneration systems. First and probably foremost is rising electrical costs. In the 1970s and into the 80s, rising fuel, materials and labor costs have dramatically increased electrical costs. Further affecting electric costs has been the high cost of bringing new plants on line. Inflation has affected construction and financial carrying costs for those utilities engaged in a construction program. In some instances, completion of a new plant and inclusion of the entire cost of the plant in rate base at one time could result in rate increases of at least 30 percent and probably much more.

A second factor sparking interest is the Public Utilities Regulatory Act of 1978 (PURPA). Essentially, PURPA was intended to encourage industrial use of cogeneration through rate incentives. A group of utilities however, challenged PURPA and the subsequent implementing rules of FERC regarding avoided cost payments to qualifying cogenerators and interconnection requirements. After a series of court battles, the U.S. Supreme Court upheld FERC's rulings in 1982. While the Supreme Court's decision has alleviated uncertainty about cogeneration at the Federal level, the battle over avoided costs has shifted to the state level. Decisions are now being made at this level between cogenerators, utilities, public utility commissions, and even the courts as to the proper contract rates and requirements for qualified cogenerators.[5]

The third factor affecting cogeneration activity is technology. Recent developments have been made both to increase efficiencies of equipment as well as to provide smaller equipment which is sized closer to most industry needs and at an economical price. These actions will increase market penetration. In a recent *New York Times* article (June 10, 1984), it was estimated that cogeneration will increase from 5% of total U.S. electricity generated in 1983 to 15% by 2000, a three-fold increase. Sales for cogeneration equipment will increase to almost $5 billion/year in 2000 as contrasted to under $½ billion in 1980.[4] Technological developments will indeed make cogeneration available to more consumers.

The fourth factor to impact on cogeneration is the future of the U.S. electric utility industry. Stories of utilities in financial difficulties because of the costs of building large central power plants are frequent. Clouding the financial issues are the uncertainties of acid rain legislation and the future of nuclear power plants. Utilities are reluctant to begin new plants. Instead, utilities are espousing conservation, load management techniques and alternate power supplies as a means to meet future needs.

AMERICAN-STANDARD'S CONCERNS

Of the four (4) factors listed above, we are concerned primarily with the first and fourth factors, i.e., rising electrical costs and generator capacity reserve margins. For the past few years, we have been involved in rate case proceedings for both electric and natural gas utilities. One dominant fact that has come out, in most if not all rate cases, is that of interclass subsidization with industry carrying the burden. Public Utility Commissions and utilities have recognized this disparity. But in my opinion, rather than correct the problem by correctly allocating costs, these bodies are taking a far more onerous position. With the encouragement of Public Utility Commissions and others, utilities are instead utilizing other cost-of-service methodologies which mathematically reduce the industrial subsidization to zero or even below. For example, in one state a cost-of-service methodology is being proposed essentially to allocate fixed generation and transmission plants on an hourly usage pattern, i.e., a kWh basis. Adoption of such a method will increase industrial electrical costs immediately by 25%, all other things being equal. Despite our efforts to interject some fairness in the development of rates, we don't believe we will be successful in the long run, unless there is a major rethinking by the political bodies involved.

We also have a concern that generator capacity reserve margins may not be as strong as claimed. While we admit readily that reserve margins are generally high today, most electrical utilities have ceased or are winding down their construction programs. Most utility executives would be considered foolish to undertake a major construction program today. Such questions as (1), Would the new plant ever be finished? (2) If it is cancelled, who would pay the sunk costs? and (3) If finished, how would the new plant be treated in rate base? are unknowns.

Figure 4-1 shows electric supply and demand through 1993:

Figure 4-1

U.S. ELECTRIC

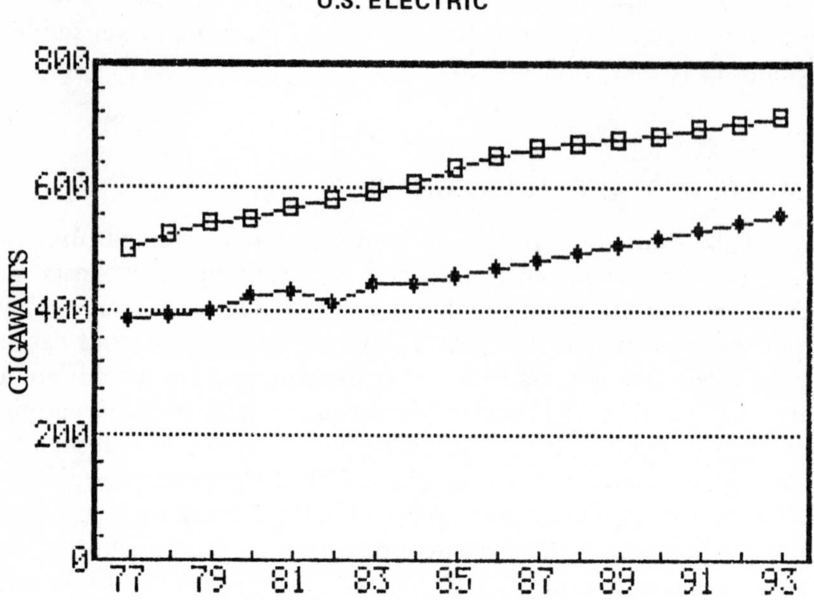

Figure 4-2 shows generating reserve margins dropping below the 30% level in the early 1900s.

The data presented in these two figures are based on utility-supplied information. While it appears that adequate reserve margins exist, we see several shortcomings. First, included in the supply or capacity figures are several plants which have been cancelled, such as Marble Hill; delayed further, such as Zimmer (proposed conversion from nuclear to coal); and subject to potential cancellation or further delays, such as Seabrook. In other words, the plants may not be there when planned. Further affecting supply may be forced outages due to maintenance and reduced capacity of units due to age and/or EPA requirements.

On the demand side, utilities have assumed an average annual growth of 2.4% per year. While this may be an accurate assumption

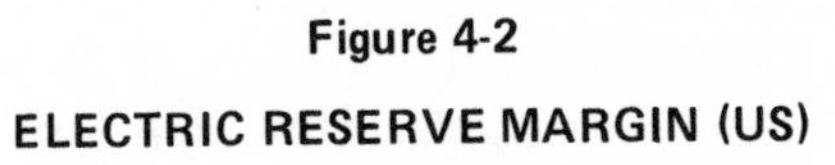

Figure 4-2

ELECTRIC RESERVE MARGIN (US)

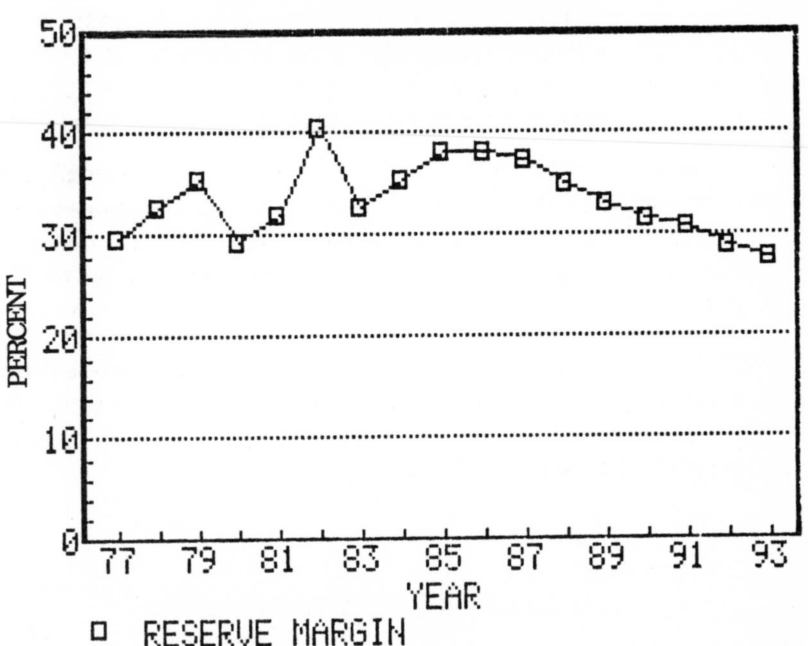

nationwide, it would be folly to assume it holds true in all regions. For example, the Sun Belt areas have experienced large amounts of growth in recent years and a continued demand growth could cause supply problems in this region and potentially in individual utilities. As a result, our cogeneration planning process is looking at individual utilities for potential supply problems.

COGENERATION TECHNOLOGY

Basically, there are three types of cogeneration systems:

(1) *Steam Turbines* – High-pressure steam is generated in a boiler and then is piped to a non-condensing steam turbine to generate electricity. The extracted steam is used in the plant for process and/or heating at a suitable pressure. Fuel for the boilers can be biomass coal, fuel oil or natural gas. Capital costs for this system can be high, but operating costs, if biomass or coal is used, can be low.

(2) *Diesel Engine Generator* – Uses a reciprocating internal combustion engine burning fuel oil or natural gas. Advantages to this system are low capital requirements and high cogeneration efficiencies. Reliability, maintainability and operating costs are a negative. Cost and supplies of natural gas fuel oils are key negative factors in using diesel generators.

(3) *Gas Turbine Generator* – Uses a combustion turbine/generator burning a light fuel oil or natural gas. This system has a low capital requirement as well as excellent reliability and maintainability. Operating costs, however, can be high because of fuel requirements and efficiencies.

We have examined all three technologies and have concluded that the gas turbine generator system best meets our needs. Essentially, we value the reasonably low capital costs coupled with the high operating reliability and maintainability.

CONSIDERATIONS IN A GAS TURBINE GENERATOR SYSTEM

Figure 4-3 is a basic schematic of a gas turbine cogeneration system:

Figure 4-3

TYPICAL COGENERATION SYSTEM

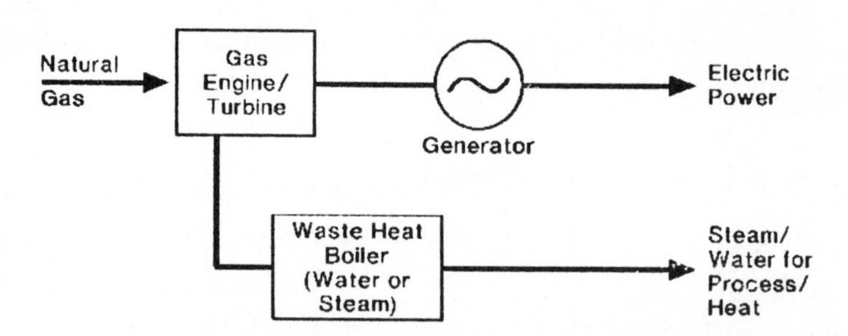

The ideal situation for such a system would be (1), a low-cost dependable gas supply, (2), high electrical costs, and (3), a good balance between thermal and electrical loads. We recognize that we do not live in an ideal world. Our facilities, like most industrial plants, do not have reliable low-cost gas supplies. Additionally, our facilities have higher electrical than thermal loads. Consequently, to overcome these drawbacks we are developing private gas gathering systems at key facilities to provide low-cost reliable fuel for our cogeneration units.

Additionally, we are reviewing methods to obtain better electrical/thermal balance through use of absorption chillers, steam use as shaft horsepower such as steam turbines in lieu of electric motors, and combined cycle generation.

SUMMARY

Cogeneration is the sequential generation of electrical and thermal power using one fuel source. It is a proven technology, previously having prominence early in the 1900s. Today, cogeneration is espoused as one of many alternatives to future capacity additions. Our studies show the importance of (1), a low-cost reliable fuel source, (2), the balance between electrical and thermal needs, and (3), the importance of reviewing the local electrical utility's cost and reserve margins. All factors are important in the successful installation of a cogeneration project.

REFERENCES

[1] AGA Manual, Cogeneration Feasibility Analysis, April 1982.

[2] "Are Utilities Obsolete?," *Business Week,* May 21, 1984.

[3] "Cogeneration – A Technology Reborn," *Industry Week,* January 23, 1984.

[4] "Cogeneration Jars the Power Industry," *New York Times,* June 10, 1984.

[5] "Cogeneration Potential Projected to Double in Capacity by 2000, Energy Analysts Project," *Energy Users Report,* p. 474, May 31, 1984.

[6] "Guidelines for Assessing the Feasibility Of Small Cogeneration Systems," a paper presented by Macauley Whiting, Jr., Vice President Decker Energy International Inc. at the 6th AEE Energy Audit and Management Symposium, June 1984.

[7] *Industrial and Commercial Cogeneration,* Office of Technology Assessment, February 1983.

[8] "Power Industry's Uncertainty," *New York Times,* May 22, 1984.

CHAPTER 5

Financing Cogeneration Projects

Dilip R. Limaye

INTRODUCTION

Cogeneration projects generally involve investment in capital equipment to produce electricity and thermal energy. Such investments produce revenues from the sale of electricity and thermal energy, or reduce the operating costs for the purchase of such energy. The financing of a cogeneration project is therefore similar to the financing of any capital-intensive project. The revenues from the sale of products from the cogeneration project (or the cost reduction resulting from the use of these products) must be sufficient to provide for the operating expenses and produce sufficient returns for the investments made in the project by the financing sources. The availability of a number of depreciation and tax benefits for cogeneration projects provides many different financing options and, in the past several years, a number of innovative financing schemes have been developed for implementing cogeneration projects.

This chapter discusses the major considerations in financing a cogeneration project, and provides an overview of financing sources and types of financing, major financing risks, and methods to minimize or contain such risks.

GENERAL CONSIDERATIONS IN FINANCING COGENERATION PROJECTS

While cogeneration projects exhibit a tremendous diversity in project size, scope, ownership, financing method and amount of financing required, the following issues must be considered in developing the optimum financing approach.

- *Credit Worthiness* – What is the credit rating of the various parties involved in the cogeneration project, and what is their capability to generate the necessary financing?
- *Type of Financing* – What are the different possible methods for financing the project, and what could be the roles of the various parties in developing the financing?
- *Financing Sources* – What are the major sources for the funds required for the cogeneration project, and what portion of the total funds will come from each source?
- *Risks and Risk Sharing* – What are the major financial risks presented by the cogeneration project, and who will share the risks and in what proportion?
- *Ability to Utilize Tax Benefits* – Can the project participants take advantage of the available tax benefits, and what is the optimal allocation of the tax benefits to the various participants?
- *Requirements of Financing Sources* – What requirements will the major financing sources impose upon the project, and how can these be best satisfied?

It is important for project developers to pay adequate attention to all of these issues in order to develop the financing plan that will provide adequate resources for the project and sufficient sharing of risks and benefits to the various project participants.

CREDIT WORTHINESS

Financing sources will always consider the credit rating of the borrower in determining how much financing to provide and under what terms and conditions. Cogeneration project developers with sufficient credit ratings (such as large corporations with a healthy balance sheet) will have no difficulties in obtaining financing for cogeneration projects. Credit-supported financing, also known as recourse financing, is quite feasible when the project developer or sponsor accepts a direct legal obligation for interest payments and debt repayments to the financing organizations. In such recourse financing approaches, the lenders will have access to some or all of the sponsor's assets outside the cogeneration project, in case the project revenues are insufficient to meet the debt and interest payments. Examples of such financing approaches include projects directly supported by large corporations and take-or-pay contracts secured by the general resources of the companies involved in the project development. Such methods are appropriate for organizations with substantial assets or other resources.

Most financial managers, however, would prefer non-recourse or project financing. In such financing methods, which are also known as off-balance sheet financing, the interest and debt payments are supported only by the cash flows from the project. The project debt is secured by a mortgage lien on the project facilities alone. None of the project sponsors or developers has a direct legal obligation to repay the debt or interest payments if the project cash flows are insufficient. The advantage of such financing to the project sponsors or developers is that their other assets or resources are not committed or at risk, and the debt does not appear on the balance sheet. From the point of view of lenders, such financing is feasible only when the basic economics of the project are attractive enough so that financing risks are minimized. In such financing methods, the lenders are more concerned with the project revenues, expenses and cash flows rather than the credit rating of the project sponsors or developers. Project financing approaches are increasingly being adopted for cogeneration projects in view of their inherent attractiveness to project developers/sponsors.

TYPES OF FINANCING

Corporate sponsors of industrial cogeneration projects often finance such projects using equity or other internal funds. In such situations, the corporate retained earnings, and/or short-term borrowings will typically finance project expenditures until the aggregate amount of the financing required for all projects is large enough to require more permanent financing, which may come from long-term debt, issuance of new equity stock, or other basic financing methods for corporations. To qualify for equity financing, the project generally must satisfy a number of corporate criteria, and must compete internally with other demands on the corporation's resources. Such internal or equity financing is adopted by corporations when:

- the corporation has access to sufficient capital at a reasonably low cost
- the project economics meet or exceed the corporate investment criteria, which are generally expressed in terms of internal rate of return or payback
- the project risks are within the risk guidelines established by the corporation
- the corporation can take advantage of the tax benefits from the cogeneration project
- external financing does not provide any significant advantages.

Because of the inherent economic attractiveness of many cogeneration projects and the ability to use project financing, many project sponsors, including some large corporations, are increasingly turning towards external financing. Such external financing methods may include one or more of the following:

- conventional lenders such as banks and insurance companies, who will finance cogeneration projects because of the returns offered by such projects
- financing by vendors of various equipment used in the cogeneration project
- third-party developers who are able to take advantage of the tax benefits and are increasingly interested in the returns offered by cogeneration
- various types of joint ventures between industrial firms, third-party developers, utilities, equipment manufacturers and other organizations – these joint ventures may be structured as new corporations or as partnerships
- limited partnerships in which individual investors provide the financing

- various types of leasing arrangements including:
  - operating lease
  - finance lease
  - safe harbor lease
  - leveraged lease

  Leasing arrangements are discussed in greater detail below.

FINANCING SOURCES

Cogeneration project developers will find that a number of different sources are available for financing their projects. Projects may be financed exclusively from one of these sources, or through a combination of different sources. It is also important to recognize that some of these financing sources will provide equity financing as well as debt to the project. An evaluation of which financing sources would be the most appropriate for a particular project depends on the type of financing commitments, the terms of financing, the size of financing required, and the approach of the financing organization to risk containment and risk sharing. A discussion of some of the common financing sources is given below.

1. Commercial Banks

Commercial banks are likely to be one of the most common sources for financing cogeneration projects. In particular, medium to large commercial banks are increasingly interested in alternative energy projects and are now convinced that cogeneration projects are attractive for conventional bank financing. For large projects, banks may form a consortium to develop the needed financing. The terms of financing from commercial banks can vary significantly from project to project, and may include fixed or variable interest rates, repayment terms ranging from 5 to 15 years, project security requirements ranging from project equipment to other assets of the developer or sponsor, and various types of guarantees or contractual requirements. Commercial banks may also be sources for construction financing during the project construction stage before permanent financing is obtained, as well as for the permanent financing after construction is complete.

2. Insurance Companies

Insurance companies have traditionally been excellent sources of financing capital projects and are increasingly interested in cogeneration. Insurance companies are generally willing to provide longer terms for financing, ranging from 15 to 20 years. They may be willing to take equity positions in certain projects, in addition to providing debt financing. Both fixed and variable rate financing may be available, and other variations in financing are quite possible. Insurance companies are also likely to be willing to take somewhat higher risks than conventional banks, as long as appropriate rewards are available to them to cover the risks taken.

3. Pension Funds

Pension funds have traditionally not been involved in energy projects, but the recent publicity for cogeneration is drawing them into the financing markets. Generally, pension funds will be willing to consider only larger projects over $5 million, and their risk/reward preferences are similar to those of insurance companies.

4. Third-Party Financing

A number of third-party financing companies have been established for the explicit purpose of financing cogeneration projects. These companies have access to a number of different financing sources and are actively searching for cogeneration projects with the right types of economic returns. Third-party financing has ranged from less than $1 million to as much as $100-200 million dollars for some projects. Quite often, these third parties have technical knowledge of cogeneration, as well as expertise in the institutional aspects of cogeneration, particularly with respect to state and federal regulations and contract negotiations with utilities and other parties. Also, these developers can take advantage of the available tax benefits and are able to pass some of the benefits to the project sponsors. Third-party financing companies are willing and interested in equity positions in projects, and there have been examples where third parties have entirely owned and operated a project, selling the project outputs to an industrial firm and/or a utility.

These types of developers are quite willing to structure innovative financial arrangements that may benefit all of the parties involved in the cogeneration project. During the last several years, an increasing number of large organizations are expressing interest in third-party financing, and are actively searching for opportunities to finance cogeneration projects.

5. Individual Investors

Because of a number of tax benefits offered by cogeneration projects, individual investors also represent a potential financing source. Limited partnerships can be structured to provide tax benefits, cash flows and potential long-term value to individual investors. While individual investors will generally require a greater return on their equity, they are more willing to take greater risks and are also willing to receive their returns over the long term, as long as some immediate tax benefits are available in the short term.

6. Equipment Manufacturers

Manufacturers of cogeneration equipment such as boilers, turbines, generators, etc., are increasingly looking to the cogeneration market as an attractive market for the sale of their products. They have recognized that being able to finance such equipment may increase their chances of making the sale to project developers or sponsors. Manufacturers are therefore increasingly providing financing to cogeneration projects. Some manufacturers have even set up cogeneration financing companies to specifically offer equity and debt financing to cogeneration projects. In the process, they have become "full service" companies that will provide a range of services including feasibility analysis, design, financing, construction, operation and maintenance.

7. Architect/Engineering Firms

A number of large architect/engineering firms, who have traditionally provided services to central station power generation markets, are now turning towards the cogeneration market. They too have recognized that their chances of making a sale increase if they offer other services such as financing. These types of organizations are willing to take equity positions in projects and are interested in providing a number of different services beyond design and construction management.

8. Industrial Development Authorities

For cogeneration projects owned by municipal or public authorities, tax exempt financing through industrial development bonds may be feasible. The advantage of such financing is that the debt cost may be reduced substantially over conventional financing methods. The ability to qualify for tax exempt financing depends on the satisfaction of a number of complex state and federal regulations.

TAX BENEFITS FOR COGENERATION PROJECTS

Since cogeneration projects offer benefits to society in terms of increased efficiency of energy utilization and reduced environmental impacts, federal and state government agencies have elected to provide a number of tax benefits to cogeneration projects. Many of the tax laws have changed over time and new tax initiatives are constantly being proposed by the Congress. Currently, there are a number of tax benefits that make cogeneration projects quite attractive. A summary of these is presented below.

As far as Federal income tax laws are concerned, there are four major tax acts that influence the benefits derived from cogeneration projects. These are:

- The Energy Tax Act of 1978 (ETA)
- The Crude Oil Windfall Project Tax Act of 1978 (COWPTA)
- The Economic Recovery Tax Act of 1981 (ERTA)
- The Tax Equity and Fiscal Responsibility Act of 1982 (TEFRA).

The ETA originated the energy tax credit for projects that reduced the nation's dependence on oil and gas. Under this act, all cogeneration projects were eligible for a 10% energy tax credit through the end of 1982, and biomass cogeneration projects were eligible for this 10% credit through the end of 1985. The COWPTA extended the types of projects eligible for the energy tax credit and the time limit for the eligibility for such credits. The ERTA offered significant depreciation benefits through the establishment of the Accelerated Cost Recovery System (ACRS). According to ACRS, most cogeneration projects are eligible for five-year depreciation. ERTA also established the safe harbor lease, whereby tax benefits could be transferred from one party to another. The TEFRA made certain changes in the depreciation and tax laws established by ERTA, and reduced the broad applicability of safe harbor leasing.

The major tax benefits that have made a number of cogeneration projects very attractive include the following:

- The business investment tax credit available to many capital investments is also available to cogeneration projects. This credit is equal to 10% of the qualified expenditures and can be claimed as an offset against the income tax liability.
- Energy tax credit available for biomass property is equal to 10% of the qualified expenditures (but expires at the end of December 1985).
- Accelerated depreciation, under the ACRS, provides a five-year write-off for cogeneration projects.
- Various local tax benefits, such as tax credits and deductions are offered by different states.

Table 5-1 provides an example of the impact of some of these tax benefits on the economics of the cogeneration project. The example assumes a cogeneration project costing a total of $10 million. It is assumed that the project developer makes an initial investment of $2 million and borrows the remaining $8 million required for the project from a conventional financing source at a 12% annual interest rate. If we assumed that the project is not a biomass cogeneration project, and therefore not eligible for the energy tax credit, the only tax credit available is the basic investment tax credit, which is 10% of the capital investment, or $1 million. According to the provisions of TEFRA, if a 10% investment tax credit is claimed, then half of this tax credit must be deducted from the project costs before depreciation is taken. The depreciation basis for the project is therefore $9.5 million dollars. Using the 5-year ACRS depreciation scheme, the first year depreciation for this project is $1.425 million. The interest cost for the project at 12% on $8 million is $0.96 million,

Table 5-1

ILLUSTRATIVE EXAMPLE – TAX BENEFITS OF COGENERATION

ASSUMPTIONS

- Project Cost – $10 million
- Investment – $ 2 million
- Debt – $ 8 million
- Interest Rate – 12 %

CALCULATIONS

- Investment Tax Credit (ITC)
  @ 10% of $10 million = $1 million
- First Year Depreciation
  @ 15% (using ACRS) of Initial Cost
  less half ITC = $1.425 million
- Interest Deduction
  @ 12% of $8 million = $0.96 million

TAX BENEFITS
(assuming 46% marginal tax bracket)

- ITC – $1,000,000
- Depreciation Benefit* – 655,500
- Interest Benefit* – 441,600

  Total Tax Benefit in First Year – $2,097,100

*calculated as 46% of deduction

leading to the cash flows for the first year shown in Table 5-1.

Assuming that the developer is in the 46% marginal tax bracket, the total tax benefits during the first year of the project (including the investment tax credit, the depreciation benefit and interest benefit) are $2.09 million. This means that the project developer recoups his entire $2 million investment within the first year. As long as the project cash flows are sufficient to pay the interest payments and debt repayment on the $8 million financing, the project represents a very attractive economic investment. This example of tax benefits illustrates the reasons why many third-party organizations who can take advantage of the tax benefits are interested in financing cogeneration projects.

ILLUSTRATIVE FINANCING STRUCTURES

A brief discussion of leasing, limited partnerships and tax exempt financing methods is provided below.

1. Leasing

Various leasing methods have proven popular for financing cogeneration projects. Leasing simply involves financing by one organization that owns the equipment or project, which is then operated by another organization with the payment of a lease fee for a specified period of time. Among the different leasing options are:

- operating lease
- finance lease
- safe harbor lease
- leveraged lease.

Each of these is briefly discussed below.

- *Operating Lease*

  This is the simplest form of leasing arrangement in which the equipment user merely pays for the use of the equipment over the specified period at a specified cost with no commitment or option to purchase the equipment at the end of the lease period. The ownership of the equipment resides with the lessor, and such a transaction is treated as a "true lease" for tax purposes.

- *Safe Harbor Lease*

  The Economic Recovery Tax Act (ERTA) created a special form of lease known as the safe harbor lease. In a safe harbor lease, the lessee arranges the acquisition of the project as well as the financing of the project, but then transfers the tax ownership to a lessor who can take advantage of the property.

In this transaction, the legal ownership stays with the lessee, even though tax ownership is transferred to the lessor. Essentially, this transaction involves a sale of the tax benefit for the equipment from the lessee to the lessor. This transaction was particularly attractive to organizations that could not take advantage of the tax benefits of cogeneration. The advantages of this type of leasing were significantly reduced with the passage of the TEFRA legislation.

- *Finance Lease*

  A finance lease is a transaction that is written as a lease but is very similar to an installment sale. In such a lease, there may be an option for the purchase of property for a specified amount at the end of the lease period, and the lessee may build equity into the property over a period of time. For tax purposes, the lessee is regarded as the owner, and is therefore entitled to take the appropriate depreciation and tax benefits. This type of lease is suitable when the lessee is in a position to benefit from the tax regulations.

- *Leveraged Lease*

  The leveraged lease works on the concept that a lessor whose capital cost is low, and who can take advantage of the tax benefits, finances the purchase of the equipment and leases it to the user of the property. The payments under the lease are lower than other forms of leases because the lessor passes through some of the tax benefits in the form of lower lease payments. In order to qualify for tax benefits, the leveraged lease must satisfy certain minimum requirements as follows:

  - the lessor must have at least a 20% at risk investment
  - the lessee cannot purchase the equipment at the end of the lease term for less than the fair market value
  - all tax benefits go to the lessor.

  Leveraged lease transactions offer advantages both to the lessor and the lessee, and have been used for financing some large cogeneration projects. Typical terms run from 5 to 10 years.

2. Limited Partnership

A limited partnership involves the use of capital available from individual investors who can take advantage of tax benefits. In a limited partnership, the major risks as well as the benefits of a cogeneration project are assigned to the individual investors. The advantage of such an arrangement is the ability to raise substantial amounts of capital with a low risk to the project developer, since the individual limited partners are generally willing to accept much higher risks than other financing sources. A limited partnership also offers the advantage that it can be structured in a number of different ways so that various options are available for allocation of the tax and other financial benefits of the cogeneration project. Generally, limited partnerships are highly leveraged with at least 60 to 80% of the financing coming from debt. The individual investors get a substantial portion of the tax benefits in the early years of the project and may benefit from additional cash flows in later years.

The disadvantages of limited partnerships relate to the high cost of identifying and organizing individual equity investors, and in structuring the appropriate legal documents for the partnership. In addition, a limited partnership must satisfy some very stringent federal and state security regulations. Therefore, such transactions require much longer lead times than other financing options. There is also a significant front-end cost involved in structuring or organizing the limited partnerships. Therefore, this option is not generally used for small projects.

3. Joint Venture

A joint venture involves the participation of multiple organizations who share the costs, risks and benefits from a cogeneration project. A joint venture is formed when the skills and expertise of the different parties are synergistic. For example, a project developer may have the technical skills to design and structure a cogeneration project, but may require help from another corporation that has the financing and regulatory expertise to make the project implementation successful. Another example of a joint venture is the collaboration of an industrial firm and a utility. In such an arrangement, the utility's ownership is restricted by the PURPA regulations to a maximum of 50%. Many industrial firms are quite interested in collaborations with a utility because the utility offers expertise in the operation of a power generation facility, and may also bring capital at a relatively low cost. Further, the participation of a utility as a part owner in a project may facilitate the negotiation of contracts for the sale of power.

The sharing of the costs, risks and benefits in a joint venture may vary from project to project. Joint ventures may be structured as a partnership or a separate corporation with equity ownership by participants.

REQUIREMENTS OF FINANCING SOURCES

The cogeneration project developer or sponsor must be aware of the major requirements of the financing sources. The most obvious and the most critical requirement of a financing source is that the basic project be economically viable. In other words, the project must generate sufficient cash flow to pay all of the project expenses, plus the interest and debt payments, plus a return on the investment of the equity participants. A second major requirement is that the project risks be minimized. Most financing organizations want a reasonable assurance that the project interest and debt payments will be met in the future. Cogeneration projects involve a number of different risks (as discussed below), and if these risks are considered to be significant, then the financing organization will insist on recourse financing requiring the project sponsor to commit his other assets or resources as collateral for the project in case of default on the debt payments. In order to minimize the risks and provide protection against long-term uncertainties, financing organizations will generally require long-term protection. In other words, it will be important for the developer to obtain contracts for his purchase of fuel, as well as for the sale of the thermal and electric energy, such that the long-term prices and costs are bounded within certain limits. If such contract protection cannot be obtained, the financing sources may consider the project inherently risky, and may require other types of protection to assure that the interest and debt repayments will be met by the project developers.

In risky situations, the financing sources will look for a debt coverage ratio on the order of 200%, i.e., the anticipated net cash flows (project revenues minus all project expenses) should be twice the anticipated interest payments and debt repayments. If the anticipated revenues and costs are guaranteed through contract protection, then the debt coverage ratio can be smaller.

A final requirement for financing sources is the sharing of benefits proportional to the risks. The higher the risk transferred to the financing source, the greater the returns the source is going to expect to receive from the project. Therefore, projects involving substantial risks may elect to obtain equity participation by various types of organizations or to establish limited partnerships with investments by individual investors who are willing to take higher risks.

RISKS AND RISK CONTAINMENT

A cogeneration project is subject to a number of different types of risks. Some of these are summarized below.

- *Design/technology* – Will the proposed design and technology perform as anticipated at the appropriate efficiency levels?
- *Construction costs* – Will the project be constructed within the estimated costs?
- *Timing/completion* – Is the project subject to significant delays or will it be completed within the schedule as anticipated?
- *Fuel supply and cost* – Will the required fuel supply be available in sufficient quantities and without significant escalation in cost?
- *Sale prices and quantities* – Will there be an adequate market for the sale of cogeneration products, and will the market accept the prices that are anticipated?
- *System performance* – Will the various pieces of equipment in a cogeneration project perform as anticipated, and produce the required outputs for the designated level of inputs?
- *Operating costs* – Are the operating costs subject to unforeseen escalations?
- *Financing costs* – Are the financing costs controlled, or is there a risk that these costs might escalate if economic conditions change?

- *Tax regulations* – Will the anticipated tax benefits actually be available, or would changes in tax regulations remove some of these benefits?

- *Regulatory and environmental risks* – Are there any significant changes in environmental or other regulations anticipated during the life of the project that may make the project unacceptable to regulatory authorities?

While all of these risks cannot be totally eliminated, it is important to limit or contain these risks so that both the project developer and the financing organization are comfortable with the level of risk faced by each relative to the returns anticipated. Typical methods that have been used to contain the risk are summarized below.

- *Design/technology risks* – Obtain quality participants who are knowledgeable regarding cogeneration technologies and design considerations.

- *Construction costs* – Obtain insurance coverage against cost escalation, or assign fixed price contracts with the construction management firm.

- *Timing/completion risks* – Obtain insurance protection or contract protection to assure that the project is completed on time.

- *Fuel supply and costs* – Sign long-term contracts with fuel suppliers.

- *Electric and thermal sales* – Obtain long-term contract protection by signing contracts with the purchasers of electricity and thermal energy.

- *System performance* – Obtain performance guarantees from the manufacturer, or insurance coverage from appropriate insurance companies.

- *Operating costs* – Obtain recognized and reputable design engineers who are knowledgeable of the appropriate technologies being used.

- *Financing costs* – Negotiate appropriate arrangements with the financing sources so that escalations of financing costs are minimized.

- *Tax regulations* – Keep track of current trends in regulatory initiatives at the federal and state levels.
- *Environmental and other regulations* – Keep track of the appropriate regulatory changes anticipated, and obtain reputable organizations knowledgeable of environmental and other regulations relevant to the cogeneration project.

It is important to recognize in addressing the risks and risk containment methods that the requirements of the project developer or sponsor and those of the financing source may, at times, conflict. For example, the financing organization would require a long-term contract for the sales of electricity to the utilities, while from a project developer's viewpoint the cost of negotiating such a contract may be significant, both in terms of legal fees and in terms of the time required to negotiate such a contract. The negotiation of a fixed-price construction contract may increase the construction costs because the construction management organization will build in the risks of the fixed price into its pricing. Because of such potential conflicts, it is important that the project developer/sponsor work very closely with the financing organization in structuring the project and in designing the risk containment methods, so that the overall project risks are minimized at the lowest possible cost to the developer and the financing organization.

SUMMARY

In summary, the choices available for financing a cogeneration project are many and varied. The selection of the optimum financing method for a particular project will depend on a large number of different factors, some of which include the following:

- The economic viability of the project
- The availability of internal capital
- The cost of equity and debt
- The ability to use the tax benefits associated with cogeneration projects
- The credit worthiness of the project sponsor or developer

- The major project risks
- The alternative methods available for containing the risks
- The risk/return preferences of the project developer or sponsor
- The need for obtaining additional expertise through a joint venture or other arrangement
- The regulatory framework for the operation of the cogeneration project
- The anticipated changes in tax, depreciation, environmental, and other major regulations affecting the project.

Adequate attention to these factors during the early stages of project design and development will facilitate the selection of the most attractive financing method for the cogeneration project.

CHAPTER 6

Fundamentals of A Third-Party Cogeneration Project

INTRODUCTION

Small packaged cogeneration units, typically 60 to 500 kW, and large 100+ MW gas turbine installations have recently captured most of the cogeneration headlines. However, a significant institutional and industrial market exists for medium-sized (2 to 20+ MW) units.

These medium-sized projects have complexities similar to the large projects: e.g., they are field erected, require auxiliary equipment, and involve different crafts during construction. But they must be managed with the tight budget approach associated typically with small projects to produce the greatest economic benefits.

A third-party owner can provide the wide range of technical expertise required to implement such a project, thereby freeing the host company, or user, from the varied details involved in implementing the project.

Based on the experience gained through a typical experience by an energy manager, certain fundamentals have emerged for the successful implementation of third-party retrofit cogeneration projects.

PROJECT CONCEPTION

The facility involved requires a large amount of electricity and steam. Figure 6-1 shows the electrical demand and how it varies throughout the day. The electrical load also varies seasonally, increasing in the summer with high air conditioning loads. The data in

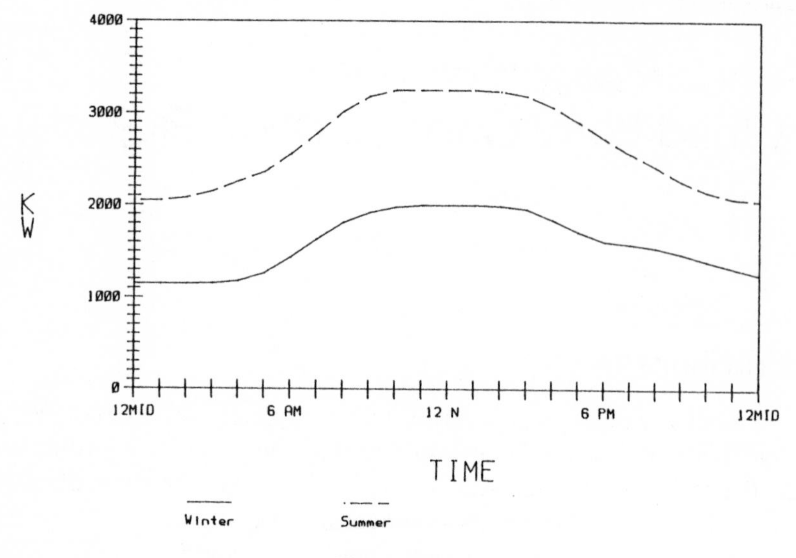

Figure 6-1

Figure 6-1 is typical for weekdays; the mid-day peaks are much less pronounced on the weekends.

Typical steam requirements are shown in Figure 6-2. The demand is greatest during the winter with increased heating requirements. In addition, the hourly changes are erratic. Domestic hot water requirements vary because the laundry operates from 8 a.m. to 4 p.m., the kitchen operates from 5 a.m. to 7 p.m., and various equipment is used imtermittently during the day.

Facility management recognized the need to hold down the energy costs to offset rising utility rates and increased usage through the addition of more advanced health care facilities. They decided to contract with the energy manager to install, own and operate a cogeneration facility and share the resulting savings in energy costs.

DESCRIPTION

The selected cogeneration facility is shown in Figure 6-3. It consists of two 1250-kW generators, each driven by a dual-fuel reciprocating engine. The engine exhausts are ducted to a waste-heat boiler

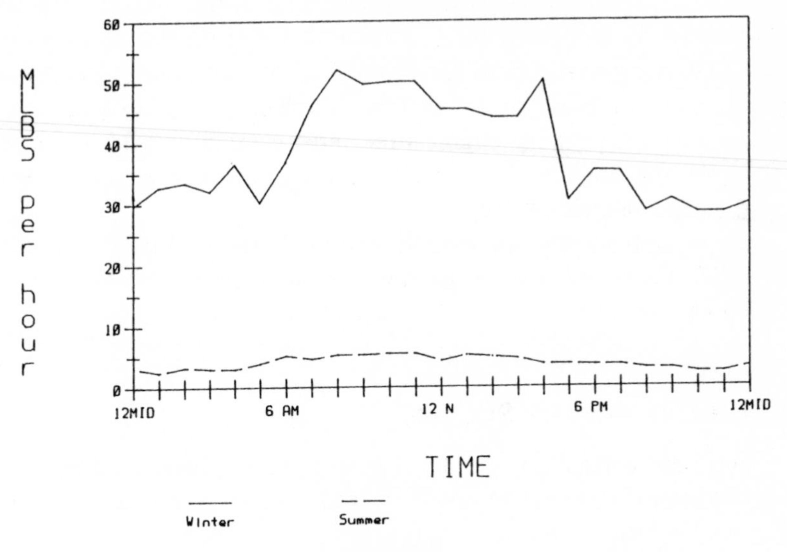

Figure 6-2

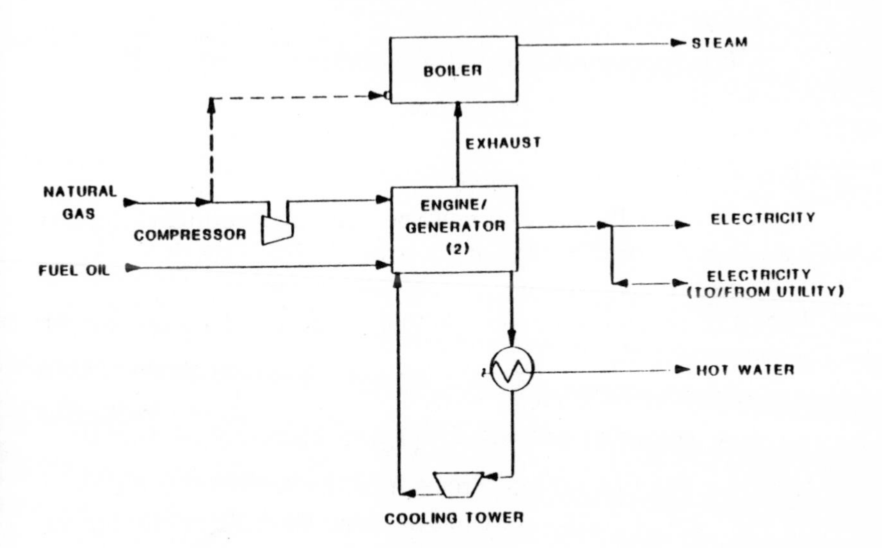

Figure 6-3

to produce 100 psig steam. The boiler can be supplementally fired to make 15,000 lb/hr of steam. Waste heat from the lube oil and jacket water is recovered in a circulating water system and used to heat the facility makeup to the domestic hot water users. Recovered heat that is not used in the facility is rejected to a cooling tower. Natural gas, the primary engine fuel, must be compressed to 55 psig for use in the engines. Excess electricity above the facility's needs is sold to the electric utility.

Figure 6-4 shows the overall energy balance. Installed cost is $1.9 MM. Projected savings to the customer are $2.9 MM over the 15-year contract life for a 6% decrease in utility costs.

REQUIREMENTS FOR SUCCESS

There are certain key elements that must be addressed during the early stages of a project to insure overall success. These key elements are:

1. Contract with the user
2. Selection of the contractor
3. Minimizing of capital cost
4. Contract with the electric utility
5. Fuel supply contract
6. Construction and operating permits
7. Operation and maintenance.

Contract with The User

The contract with the user must be structured so that each party has a stake in the profitability of the project. Cooperation is essential to the success of the venture.

Typically, the following issues should be addressed in the contract:

- Base data from which the savings will be measured. This is usually an historical average of fuel and electrical costs adjusted, if necessary, to the present operating conditions.
- Calculation of savings. Procedures for determining the revenues and expenses should be described. Revenues for the project include avoided electric energy cost, avoided electric demand charges, sale of electric to the utility, sale of steam

ENERGY BALANCE

		(1) MMBTU/HR
IN		
Natural Gas	24.2 MSCFH	22.58
Fuel Oil	8.8 gal./hr.	1.15

TOTAL		23.73
OUT		
Net Electricity	2310 KW	7.89
Steam	4200 lb./hr.	4.22
Hot Water	---	1.03
Losses	---	10.59

TOTAL		23.73
EFFICIENCY		
Electrical	33.2%	
Thermal	22.1	

TOTAL	55.3%	

(1) Net Heating Value

Figure 6-4

and hot water to the facility, and avoided operation and maintenance costs due to reduced use of existing boilers. Expenses include natural gas, diesel fuel, operating labor, maintenance costs, insurance, supplies, and debt service.

- Savings split between client and third-party owner. The savings produced by the cogeneration facility must be shared in such a way that the client receives enough to make it worth his while to enter into an agreement and the third-party owner receives a return high enough to induce him to make the investment.
- Contract length. This is typically 10 to 25 years.
- User obligations. The user should be required to utilize all the facility outputs to the fullest extent possible. Also, the user should protect the facility in the same manner that he protects his own property.

- Future modifications. The contract should state what agreement is required between the parties in case it is desired to modify the facility sometime during the life of the contract. Also, future changes in the user's operation may affect savings from the cogeneration system. Any such change should be cause for renegotiation of the contract terms to ensure that savings continue to be split equitably.
- Options for the user to purchase the facility during the term of the contract. Possibilities are an agreed-upon fair market value or a depreciated installed cost.
- Disposition of the facility at the end of the contract life. Purchase by the user at a fair market price, dismantling and removal of the facility, or operation under a new contract, are three posibilities.

Contractor Selection

This is probably the single most important factor in successfully completing the project. Great care must be taken to select a contractor who can complete the construction within the budget and schedule. The first task is to determine whether the contractor has the experience to perform the work. Some attributes to look for are:

- A good track record of satisfied customers
- Financial soundness
- Previous experience with retrofit projects
- Engineering, design and procurement capabilities
- Capable and experienced subcontractors
- Previous experience with the specific type of equipment to be installed

A lump sum contract should be executed if at all possible. The installed cost is more likely to be kept within budget. The contract should include suitable specifications, a schedule for the submittal of drawings and data, a construction schedule, and progress payments and penalties tied to certain milestones.

The contract should also include a performance test with a penalty if the unit does not pass the performance criteria.

Minimize Capital Cost

This is often essential in order to make the project economically viable. Several things can be done to hold down the initial cost.

- Install only what is essential. All items should be of good quality, but frills and niceties should be avoided. Everything should be economically justified as required for operation, safety, and/or maintenance.
- Select a contractor with experience in designing, specifying and procuring the equipment and material to be used. A well intentioned, but incompetent, contractor will certainly add to the cost, not to mention lost revenue caused by delays in completing the project.
- Used equipment. This can be a great source of cost reduction. Any used equipment should be inspected thoroughly and rebuilt as necessary to return it to the original specifications and tolerances. Properly reconditioned and maintained, used equipment will provide many years of satisfactory service.

Electric Utility

The electrical interconnection will require approval by the electric utility and, if sale of electricity to the utility is in the project scope, then a contract will have to be negotiated.

How difficult this aspect of the project will be depends a great deal on the philosophy of the utility toward cogeneration. PURPA mandates that the utility must buy any excess electricity that a qualified cogenerator wishes to sell. However, it does not dictate what requirements the utility may place on the cogenerator, nor does it set any time requirements for the utility to settle the purchase agreement.

Discussions with the utility should be started as early as possible to try to understand all the requirements. Topics that must be settled with the utility are protection requirements for the interconnection, buyback terms, availability of backup power and maintenance power at reduced costs, metering requirements, telemetry requirements, insurance and indemnification, term of the contract, maintenance requirements, and the utility's right to disconnect the facility under certain conditions.

Whether or not the project is the utility's first cogeneration effort will have a bearing on the difficulty in reaching an agreement. The utility may not have all its policies and procedures formalized for the initial project. This can lead to honest misunderstandings and delays. Also, the utility is highly aware of precedent and sometimes treats any concession as being lost forever. Therefore, the utility can be hesitant on any issues they want to preserve for the future, possibly for larger cogeneration projects.

Fuel Supply

Contact should be made with the local fuel utility or supplier as early as possible in the project. Fuel will probably be the largest expense on the cogeneration project. Usually, one contract for the supply of fuel and another for the transportation are required. Under most circumstances fuel contracts should be signed at project inception to lock in the major operating cost. However, current trends in fuel prices can affect the strategy toward finalizing the contracts. For instance, natural gas prices have been decreasing recently. We were able to arrange short-term gas at a reasonable rate and delay signing a long-term contract. This resulted in a long-term contract for gas at $0.50 per MCF cheaper than was available six months earlier.

Contract life should be identical to the life of the cogeneration shared savings contract. Other items in the fuel contract should be the means of adjusting for inflation, minimum and maximum quantities, take or pay provisions and price.

For this project it was possible to tie the inflation adjustor to the change in electric rates. This means that the fuel cost (the major cost) will vary with the electric rates (the major revenue), thereby assuring a fairly constant margin for the life of the project.
the project.

The third-party owner should be adept at sourcing fuels (e.g., gas, oil, coal, wood) and knowing prevailing market conditions in order to get the best fuel deal available.

Permits

All necessary permits must be identified early and agreement reached on which party (user, owner, contractor or outside consultant) will be responsible for obtaining each. Typical permits include FERC Qualifying Facility certification, air quality, waste water quality, building, and operating permits.

Operations and Maintenance

The third-party owner will arrange to operate and maintain the installation. Often the most logical method is to contract with the user to provide operating personnel. This is particularly true if the user can operate the cogenerating facility with his existing manpower. Operating costs to be paid to the user can be based on the amount of power produced or Btus sold to the user.

Every attempt should be made to include a maintenance contract with the installation contract. Since the contractor will have responsibility for the unit over a number of years, he will have an incentive to make sure that the installation is of high quality.

Other ongoing responsibilities, such as purchasing supplies and repairing and maintaining any items not covered by a maintenance contract, should be agreed upon and assigned as early as possible.

Monitoring of critical operating data is essential to assure that the unit is operated at maximum efficiency. This is being done on the project with a local programmable controller acquiring all data related to billing and engine operation and transmitting it via phone link to the energy manager's facility. Preprogrammed software collates the data and prints a daily report that gives hourly billing information and relationships such as efficiency of each engine, boiler efficiency, overall plant efficiency, and amount of waste heat recovered from the engines and used in the facility. Also, a monthly summary is printed out for use in billing.

CONCLUSION

Medium-sized cogeneration projects offer energy users an opportunity to reduce overall energy costs. A third-party owner and operator allows the user to realize these reduced costs without making the initial capital investment. The third-party owner, having the expertise to address project fundamentals, will assure successful implementation of the facility.

PART III

COGENERATION SYSTEMS/TECHNOLOGIES

CHAPTER 7

GAS TURBINE TECHNOLOGY

Part A:

Overview, Cycles & Thermodynamic Performance

Cyrus B. Meher-Homji and Alfred B. Focke

INTRODUCTION

Gas turbine engines are widely used in cogeneration, electric utility, gas distribution (pipeline), aircraft, and mechanical drive applications. These engines represent years of innovative development work since the early 1900s. A common thread throughout the gas turbine's evolutionary journey was the battle to obtain higher compressor efficiencies and higher cycle peak temperatures. These two goals were inextricably linked with advancements in aerodynamics and metallurgy. The industrial gas turbine has now gained acceptance as a reliable prime mover. Early gas turbines had poor cycle efficiencies (12-17%) and this somewhat hindered their acceptance, but currently thermal efficiencies in the 26-35% range are attainable. An historical trace of gas turbine development is shown schematically in Figure 7-1.

There are several factors which make gas turbines attractive:

- High Power/Weight Ratio
- Compactness
- Fast Lead Time for Installation
- Significant Fuel Flexibility

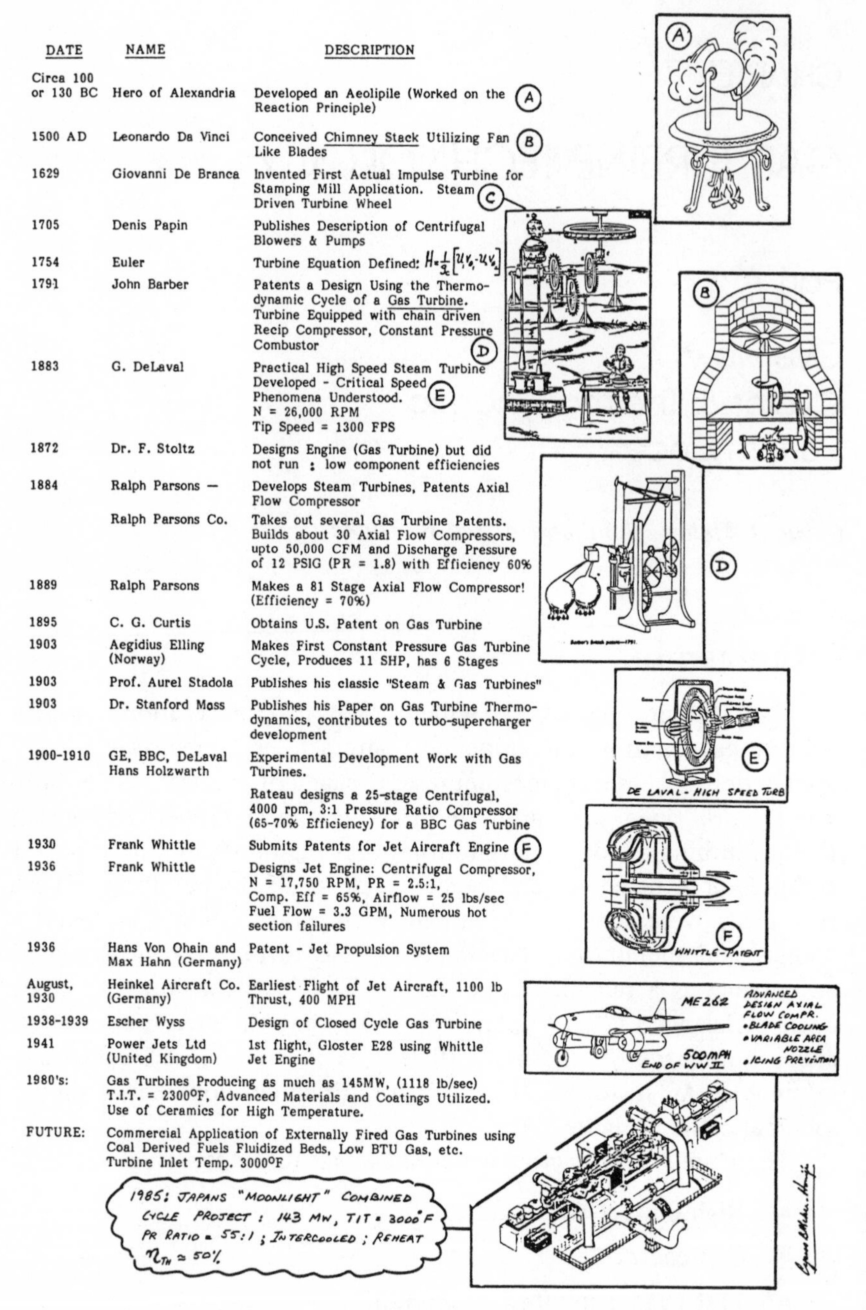

DATE	NAME	DESCRIPTION
Circa 100 or 130 BC	Hero of Alexandria	Developed an Aeolipile (Worked on the Reaction Principle) (A)
1500 AD	Leonardo Da Vinci	Conceived Chimney Stack Utilizing Fan Like Blades (B)
1629	Giovanni De Branca	Invented First Actual Impulse Turbine for Stamping Mill Application. Steam Driven Turbine Wheel (C)
1705	Denis Papin	Publishes Description of Centrifugal Blowers & Pumps
1754	Euler	Turbine Equation Defined: $H = \frac{1}{g_c}[U_1V_{u_1} - U_2V_{u_2}]$
1791	John Barber	Patents a Design Using the Thermodynamic Cycle of a Gas Turbine. Turbine Equipped with chain driven Recip Compressor, Constant Pressure Combustor (D)
1883	G. DeLaval	Practical High Speed Steam Turbine Developed - Critical Speed Phenomena Understood. (E) N = 26,000 RPM Tip Speed = 1300 FPS
1872	Dr. F. Stoltz	Designs Engine (Gas Turbine) but did not run : low component efficiencies
1884	Ralph Parsons —	Develops Steam Turbines, Patents Axial Flow Compressor
	Ralph Parsons Co.	Takes out several Gas Turbine Patents. Builds about 30 Axial Flow Compressors, upto 50,000 CFM and Discharge Pressure of 12 PSIG (PR = 1.8) with Efficiency 60%
1889	Ralph Parsons	Makes a 81 Stage Axial Flow Compressor! (Efficiency = 70%)
1895	C. G. Curtis	Obtains U.S. Patent on Gas Turbine
1903	Aegidius Elling (Norway)	Makes First Constant Pressure Gas Turbine Cycle, Produces 11 SHP, has 6 Stages
1903	Prof. Aurel Stadola	Publishes his classic "Steam & Gas Turbines"
1903	Dr. Stanford Moss	Publishes his Paper on Gas Turbine Thermodynamics, contributes to turbo-supercharger development
1900-1910	GE, BBC, DeLaval Hans Holzwarth	Experimental Development Work with Gas Turbines.
		Rateau designs a 25-stage Centrifugal, 4000 rpm, 3:1 Pressure Ratio Compressor (65-70% Efficiency) for a BBC Gas Turbine
1930	Frank Whittle	Submits Patents for Jet Aircraft Engine (F)
1936	Frank Whittle	Designs Jet Engine: Centrifugal Compressor, N = 17,750 RPM, PR = 2.5:1, Comp. Eff = 65%, Airflow = 25 lbs/sec Fuel Flow = 3.3 GPM, Numerous hot section failures
1936	Hans Von Ohain and Max Hahn (Germany)	Patent - Jet Propulsion System
August, 1930	Heinkel Aircraft Co. (Germany)	Earliest Flight of Jet Aircraft, 1100 lb Thrust, 400 MPH
1938-1939	Escher Wyss	Design of Closed Cycle Gas Turbine
1941	Power Jets Ltd (United Kingdom)	1st flight, Gloster E28 using Whittle Jet Engine
1980's:	Gas Turbines Producing as much as 145MW, (1118 lb/sec) T.I.T. = 2300°F, Advanced Materials and Coatings Utilized. Use of Ceramics for High Temperature.	
FUTURE:	Commercial Application of Externally Fired Gas Turbines using Coal Derived Fuels Fluidized Beds, Low BTU Gas, etc. Turbine Inlet Temp. 3000°F	

Figure 7-1

HISTORICAL DEVELOPMENT OF GAS TURBINE — JET ENGINE

Some important criteria to be evaluated when specifying a gas turbine are:

- Efficiency – Life Cycle Costs are made up in large part of fuel costs. The annual fuel costs on a 25MW gas turbine could easily run between 7 and 8 million dollars.
- High Reliability, thus High Availability – This is a strong function of system design, fuel used and environmental factors.
- Ease of Service – Especially good with aeroderivative engines.
- Ease of Installation and Commission–Several manufacturers provide pre-engineered modular construction allowing rapid deployment.
- Conformance with Environmental Standards – For most turbines, some sort of NO_X control has to be used.
- Incorporation of Auxiliary and Control Systems which have a high degree of reliability – Auxiliary and controls account for a significant amount of downtime. Careful specification with an emphasis on reliability is a must.
- Flexibility to meet various services and fuel needs – While gas turbines can operate on a wide variety of fuels, natural gas and light distillates are most common in cogeneration applications. Machines can be required to have dual or tri-fuel capability.

Gas turbines may be classified into two basic categories: Industrial heavy duty and the aeroderivative. Personal preferences have often created debates regarding the "preferability" of these two types. Both have their advantages and disadvantages depending on application. Technological changes are slowly bringing these two design philosophies together. Several engines on the market now "combine" the two design philosophies.

INDUSTRIAL HEAVY DUTY GAS TURBINES

These gas turbines were designed shortly after World War II and introduced to the market in the early 1950s. The early heavy-duty gas turbine design was largely an extension of steam turbine design. Restrictions of weight and space were not important factors for these ground-based units, so the design characteristics included heavy-wall casings split on horizontal centerlines, sleeve bearings, large-diameter combustors, thick airfoil sections for blades and stators, and large frontal areas. The overall pressure ratio of these units varies from 5:1 for the earlier units to 15:1 for the units in present-day service. Turbine inlet temperatures have been increased and run as high as 2000°F.

The most widely used industrial heavy-duty turbines employ axial flow compressors and turbines. In most United States designs, combustors are canannular, while some European designs utilize side combustors. As these machines have large frontal areas, inlet velocities are reduced, thus reducing noise levels.

AERODERIVATIVE GAS TURBINES

Aeroderivative engines are basically aircraft engines that have been adapted and modified for industrial use (longer life and ground based use). As the flow characteristics of the propelling nozzle (used on jet engines) and power turbines are almost identical, mechanical power can be extracted by adding a power turbine to a gas generator. The gas generator (or gas producer) unit includes the compressor, combustor and turbine(s) driving the compressor. The power turbine is coupled to the gas generator and may be designed and manufactured by a "package supplier." Aeroderivative engines have seen many applications in the cogeneration market. They are light machines and are designed for modular maintenance.

GAS TURBINE COGENERATION

The growth of cogeneration technology started in the 1960s, but has recently accelerated. It is estimated that 50% of the market will be dominated by gas turbine topping cycles. Gas turbine cogeneration is far more efficient than a typical utility central plant. About 75% heat utilization can be realized for power and heat, with about 25% leaving as exhaust. In a fossil plant, only 35% is obtained as power with condenser losses and boiler losses accounting for around

48 and 15% respectively. The gas turbine is a desirable prime mover when the power-to-heat ratios are high.

GAS TURBINE THERMODYNAMICS & CYCLES

The gas turbine is a constant volumetric flow rotating machine at a given speed. Air is ingested into the compressor and routed to the combustion section where fuel is added increasing the energy level. The hot high-pressure air is then expanded in the turbine.

Gas turbine engines utilize a thermodynamic cycle known as the Brayton Cycle which is shown in Figure 7-2 (a) on a Pressure-Volume & Temperature-Entropy diagram. In this figure, process 1-2 represents compression, 2-3 represents constant pressure heat addition in the combustor and 3-4 represents the expansion process. At point 4, the airflow is exhausted to atmosphere and a fresh supply of air is ingested into the compressor. In the case of closed-cycle gas turbines (described in a section ahead) the working fluid is cooled down to state 1 and reutilized.

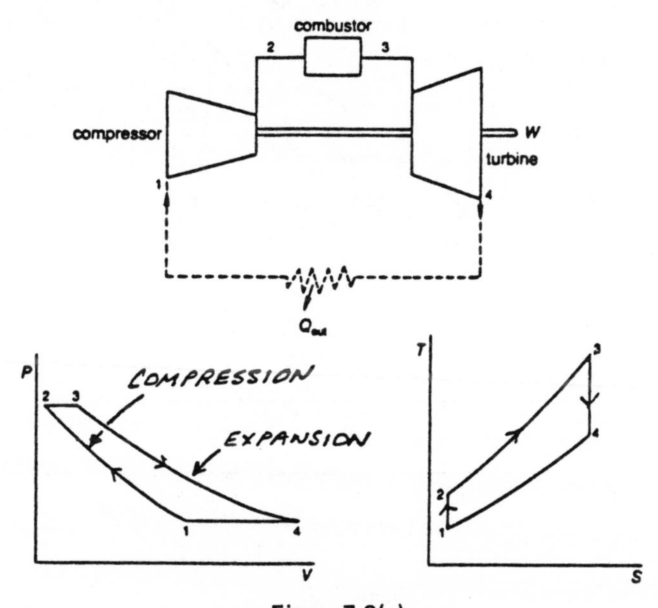

Figure 7-2(a)
PRESSURE-VOLUME & TEMPERATURE ENTROPY DIAGRAM FOR A GAS TURBINE

A combined cycle is shown on a T-S plane in Figure 7-2 (b). The gas turbine heat rejection from 4-1 is used to generate superheated steam (A-D on the Rankine Cycle) which may then be expanded in a steam turbine (D to E).

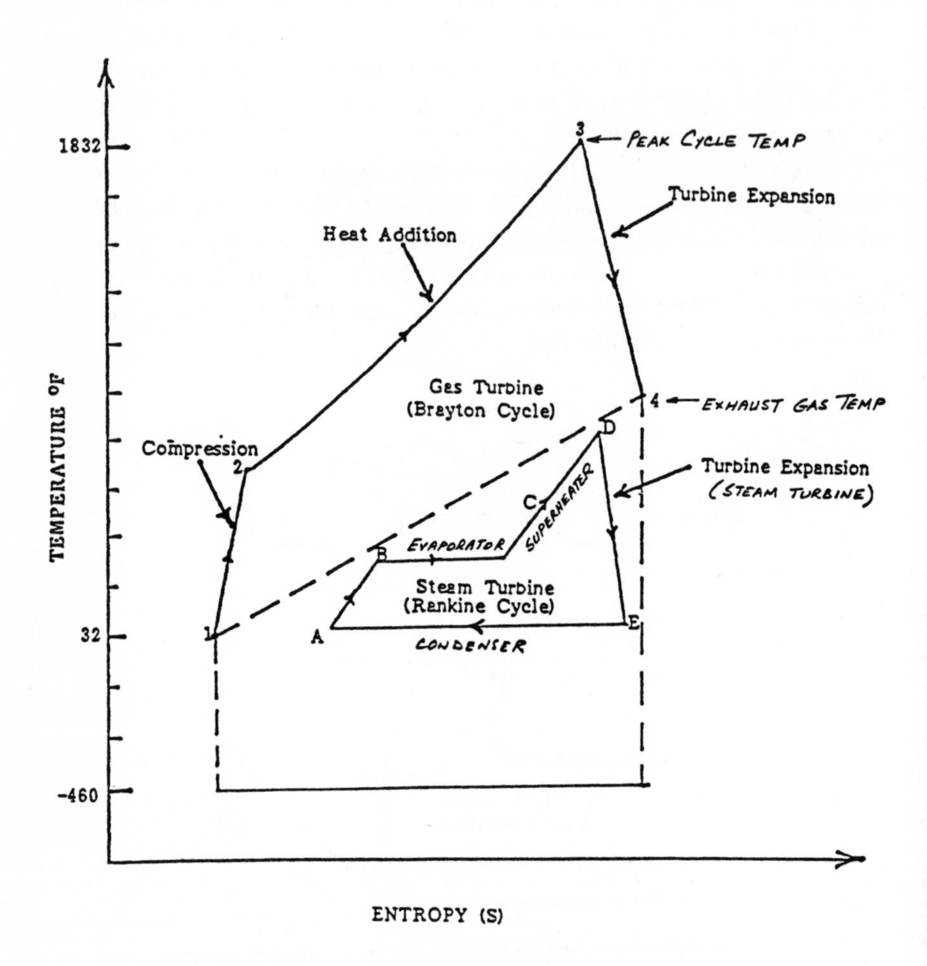

Figure 7-2(b)
BRAYTON-RANKINE COMBINED CYCLE

Compressor Section

The basic equation for adiabatic compression is:

$$\frac{T_2}{T_1} = \left(\frac{P_2}{P_1}\right)^{\frac{\gamma-1}{\gamma}}$$

where

$$\gamma = \left(\frac{C_p}{C_v}\right)$$

Assuming that the process does not involve any heat addition, the work is given by the difference in enthalpies at the compressor inlet and exit.

$$\text{Actual Work} = C_p T_2 - C_p T_1 = (h_2 - h_1) \frac{\text{BTU}}{\text{lb}}$$

The compressor efficiency is defined by:

$$\eta_c = \frac{\left[\left(\frac{P_2}{P_1}\right)^{\frac{\gamma-1}{\gamma}} - 1\right]}{\left[\frac{T_2}{T_1} - 1\right]} = \frac{(\Delta h)\ \text{Ideal}}{(\Delta h)\ \text{Actual}}$$

The compressor work is given by:

$$\text{Actual Work} = \dot{m}\, C_p \,(T_{2act} - T_1)$$

$$= \frac{\dot{m}\, C_p\, T_1}{\eta_c} \left[\left(\frac{P_2}{P_1}\right)^{\frac{\gamma-1}{\gamma}} - 1\right] \frac{\text{BTU}}{\text{SEC}}$$

Where:

$\dot{m}$ = Airflow Rate in Lbs/Sec.

The temperatures utilized here are total or stagnation values (measured by a thermometer).

$$\text{Total Temperature} = \text{Static Temperature} + \frac{V^2}{2g\ J\ C_p}$$

Where:

V = Stream Velocity

Combustor Section

In the combustor section, constant pressure heat addition occurs. The enthalpy at the burner exit is equal to the entry enthalpy (from compressor discharge) plus the enthalpy of heat addition:

$$C_p\ T_3 = C_p\ T_2 + Q$$

From the ideal heat consumption per lb of airflow is:

$$Q = C_p\ T_3 - C_p\ T_2$$

The actual wt of fuel ($\dot{m}_f$) and airflow ($\dot{m}_a$) are related by:

$$\dot{m}_f\,(LHV)\ \eta_b = (\dot{m}_a + \dot{m}_f)\ C_p\ T_3 - \dot{m}_a\ C_p\ T_2$$

Where:

η_b = Combustor Efficiency

Typically, combustor efficiencies are between 95-98% and fuel flow rates are 1-2% of airflow rate ($\dot{m}_{air}$).

Turbine Section

The heated air which could be at a firing temperature of 1700-2200°F is expanded in the turbine down to around 900-1000°F (Typical Exhaust Gas Temperature).

The turbine work is given by:

$$W_T = (\dot{m}_a + \dot{m}_f)\ C_p\ (T_3 - T_4)\ \frac{\text{BTU}}{\text{SEC}}$$

The Ideal Expansion process is defined as:

$$\frac{T_3}{T_4} = \left(\frac{P_3}{P_4}\right)^{\frac{\gamma-1}{\gamma}}$$

The Turbine Work can also be expressed as:

$$W_T = (\dot{m}_a + \dot{m}_f)\ C_p\ T_3\ \eta_t \left[1 - \frac{1}{\left(\frac{P_3}{P_4}\right)^{\frac{\gamma-1}{\gamma}}}\right]$$

The values of Cp (BTU/lb °F) should be average values between the two temperatures under consideration. Typical values are:

C_p = .246 for the Compressor Section (between T_1 and T_2)

C_p = .274 for the Turbine Section (between T_3 and T_4)

In a single shaft turbine, a significant portion (60 to 70%) of the turbine work goes into driving the compressor. The net output neglecting mechanical losses is given by:

$$W_{op} = W_T - W_C - W_{\text{Mech Losses}}$$

The overall cycle efficiency or thermal efficiency of the gas turbine is given by:

$$\eta_{th} = \frac{\text{Work Output}}{\text{Energy Input}} = \frac{W_T - W_C}{\dot{m}_f\,(\text{LHV})\ \eta_b}$$

By making certain assumptions such as negligible fuel flow compared to the airflow, constant specific heats, identical compression and expansion ratios and 100% component efficiency, the following expressions for thermal efficiency can be obtained:

$$\eta_{th} = 1 - \frac{1}{(\text{PR Ratio})^{\frac{\gamma-1}{\gamma}}} = 1 - \frac{T_4}{T_3}$$

Thus cycle efficiency may be increased by increasing the pressure ratio, increasing the turbine inlet temperature. The effect of turbine inlet temperature is predominant: A 100°F increase in turbine inlet temperature yields about a 10% increase in work and a 1½% increase in efficiency.

Gas Turbine Cycle Variations

There are several variations of the Simple cycle such as the Regenerative cycle, Steam Injected cycle, Combined (Brayton-Rankine) cycle and Reheat cycle. For practical cogeneration applications, the Simple cycle and Combined cycle are commonly used.

Figures 7-3(a) and (b) show a Performance Map for a Simple cycle and a Combined cycle.[1] These maps provide Thermal Efficiency Vs. Net Output work in BTU/lb air for varying pressure ratios and turbine inlet temperatures. It can be seen that for a specified inlet temperature T, an "optimal" pressure ratio exists above which net output/lb drops.

In several Combined cycle cogeneration schemes a condensing extraction steam turbine is used allowing swings in steam to process demand.

Single and Dual Shaft Gas Turbines

Gas turbines used for mechanical drive turbines and most aero-derivative gas turbines have a dual shaft configuration. The first shaft has the compressor and a gas generator turbine which powers the compressor. The power turbine drives the load. This configuration allows higher efficiency at part loads as the compressor section can be kept close to its design speed. Frequently, variable power turbine nozzles are provided to distribute the enthalpy drop between the gas generator and power turbine sections.

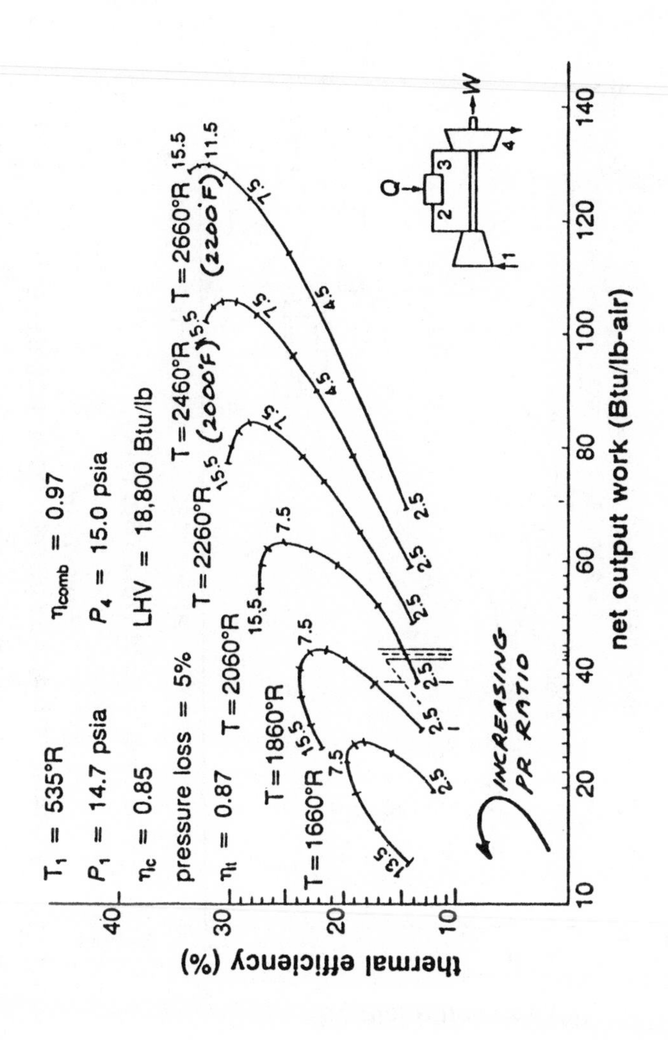

Figure 7-3(a)

PARAMETRIC ANALYSIS OF A SIMPLE CYCLE[10]

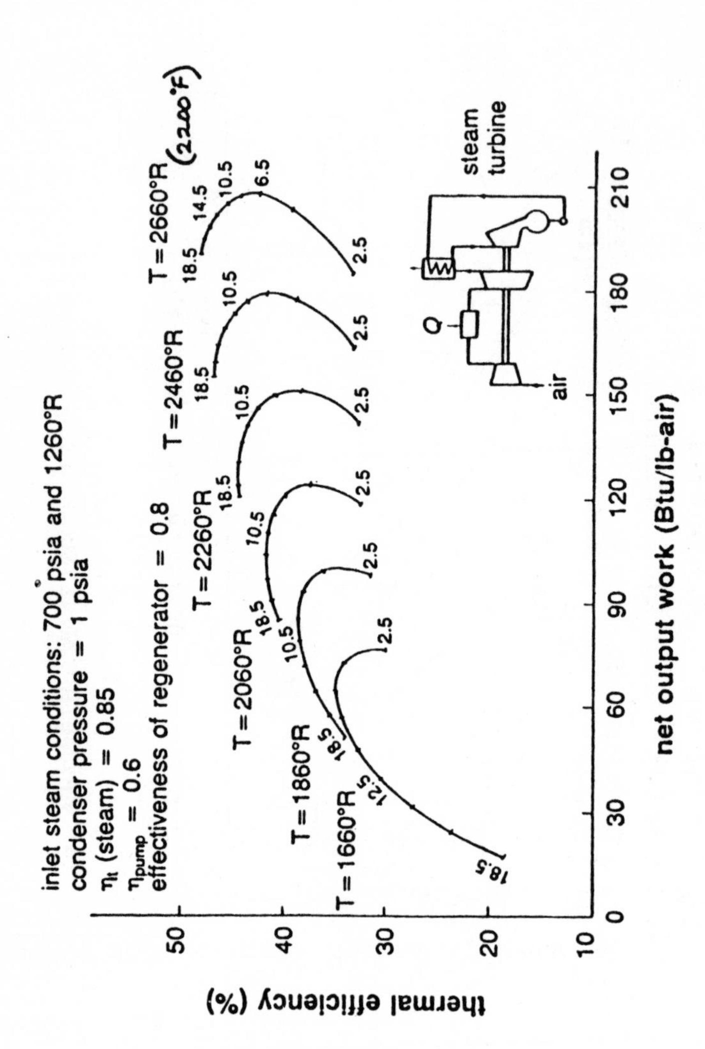

Figure 7-3(b)
PARAMETRIC ANALYSIS OF A COMBINED CYCLE[10]

Influence of Site Conditions on Gas Turbine Performance

There are several factors that influence the site performance of a gas turbine. These include:

- Ambient Temperature
- Altitude
- Inlet and Exhaust Pressure Drops

Most manufacturers provide gas turbine performance under standard conditions known as ISO conditions (59°F, 14.7 PSIA, 50% RH) and with no inlet or exhaust pressure drops.

When evaluating gas turbine performance, this must be kept in mind, along with factors such as:

1. Performance changes with different fuels: Some turbine manufacturers quote a 2-3% better performance when using natural gas.
2. Losses and Inefficiencies: There are several losses which the manufacturer may or may not include on his preliminary specification sheet. These include Inlet and Exhaust losses (described ahead), Gearing losses (can result in about a 1.5% penalty in power), Electric Generator losses (can result in a 2% drop in output and auxiliary plant power which could be 3-4% of generated power for a combined cycle cogeneration plant).

A. Ambient Temperature

A very basic problem with operation of gas turbines in hot climates is the adverse influence of high ambient temperatures on output power. The work output from a turbine can be expressed as:

$$W_O = (\dot{m}_a + \dot{m}_f)\ C_p\ (T_{in} - T_{exh})$$

Where:

W_O = Work Output

$\dot{m}_a$ = Mass Flow Rate of Air

$\dot{m}_f$ = Mass Flow Rate of Fuel

C_p = Average Specific Heat across Turbine

T_{in} = Turbine Inlet Temperature

T_{exh} = Exhaust Gas Temperature

As the airflow through the machine drops, the work output also drops. High ambient temperature causes a drop in air density and a consequent decrease in air mass flow rate through the machine. A rule of thumb is that for a 1°F increase in ambient temperature, output drops .5%.

Another effect of high ambient temperature is the decreasing of compressor pressure ratio which further decreases work output and causes the thermal efficiency to drop. Power reduction can at times be as high as 30-40%. Moreover, excessively high ambient temperatures impose severe loads on the turbine cooling systems as turbine inlet temperatures have to be boosted.

The gas turbine is a constant inlet volumetric flow rate machine for a given rotational speed. The mass flow rate is proportional to the absolute compressor inlet pressure and inversely proportional to the absolute inlet temperature.[2] Also, the mass flow rate ($\dot{m}$) is proportional to the absolute pressure at the turbine inlet nozzle and is mach number limited here. This makes it inversely proportional to the square root of the absolute turbine inlet temperature.

Mathematically, $$\dot{m} = K_1 \left(\frac{P_1}{T_1}\right) = K_2 \sqrt{\frac{P_3}{T_3}}$$

One way to recover lost power is by increasing turbine inlet temperature. This is not, however, a feasible alternative because of material limitations and the serious effect of hot section life reduction. It is interesting to note that in a typical gas turbine engine, as much as 50% to 60% of the total work produced by the turbine is used by the engine's compressor. Thus, any scheme of decreasing compressor power consumption would result in more shaft horsepower.

A review of different cooling schemes (the most common of which is evaporative cooling) is provided in reference three.[3] Manufacturers' performance maps provide variations with ambient temperatures.

B. Effects of Altitude

As the altitude increases, the pressure drops and the gas turbine experiences a drop in output. The power loss is roughly 3-4% for a thousand feet.

C. Inlet and Exit Pressure Drops

The inlet and exit pressure drops are a function of system design. Inlet pressure drops are caused by inlet duct filtration systems and silencers. The inlet ΔP acts in a similar way to an increase in altitude. A typical inlet loss is 4″ of H_2O. Exhaust back pressure occurs because of restrictions in exhaust duct and the waste heat recovery boiler and should be between 6-10″ in a cogeneration application. An increased back pressure decreases the expansion ratio of the turbine sections, thus reducing work output. A typical power loss is around .48% per inch inlet depression and .18 to .24% per inch of exhaust depression (back pressure). As rough rules of thumb, a 4″ inlet loss will cause a .7% increase in heat rate and a 2°F increase in exhaust gas temperature.

GAS TURBINE PERFORMANCE AND HEAT RECOVERY

Properties of Gas Turbine Exhaust

Exhaust from gas turbines in the simple cycle or even regenerative mode, represent a large amount of waste heat. Mass flow rates can vary depending on the size of the gas turbine from 4 lbs/sec for a 180 kW unit to a tremendous 1100 lbs/sec for a large 143 MW gas turbine. In most cases, exhaust temperatures will be in the neighborhood of 900-1000°F. As gas turbines operate with large amounts of excess air, about 18% oxygen is available in the exhaust and this allows the use of supplemental firing.

Temperatures after supplemental firing can be as high as 1600°F and it is often possible to double the steam production by supplemental firing. Variation in the exhaust gas specific heat for combustion is linear with temperatures about .259 @ 800°F to about .265 @ 900°F (C_p in Btu/lb °F).

Steam Generation Calculations

Most gas turbine manufacturers have a set of curves for their different gas turbine models showing steam generating capabilities under different conditions. Figure 7-4 shows such a curve for a 12 MW gas turbine. If curves are not available, a relatively simple calculation with regard to the waste heat boiler can be done to get a ballpark estimate. Figure 7-5 shows a schematic representation of a simple waste heat recovery boiler (WHRB). In order to compute steam generated, the following items are required:

- Gas Turbine Mass Flow Rate (lbs/hr)
- Exhaust Gas Temperature (°F)
- Steam conditions (Psig/Temp °F) required
- Feedwater Temperature (°F).

Some general rules of thumb can be stated as they are implicit in the computation. First, steam outlet temperature cannot be closer than about 50°F of the exhaust gas temperature. Also, the exhaust gas leaving the boiler evaporator must be 40°F greater than the saturation temperature. The pinchpoint is taken to be not less than 40°F and the water entering the evaporator is 15°F less than saturation temperature. Based on a simple heat balance we have:

$$\dot{m}_{steam} = \frac{\dot{m}_{air} \times C_p \times (T_2 - T_1) \times K}{(h_2 - h_1)}$$

where,

$\dot{m}_{steam}$ = Steam Flow in (lbs/hr)

$\dot{m}_{air}$ = Gas Turbine Mass Flow Rate (lbs/hr)

T_2 = Exhaust Gas Temperature

T_1 = Stack Exhaust Temperature (saturation + 40°F)

C_p = Specific Heat at Constant Pressure (average between T_1 & T_2)

h_2 = Steam Final Enthalpy (Leaving HRSG)

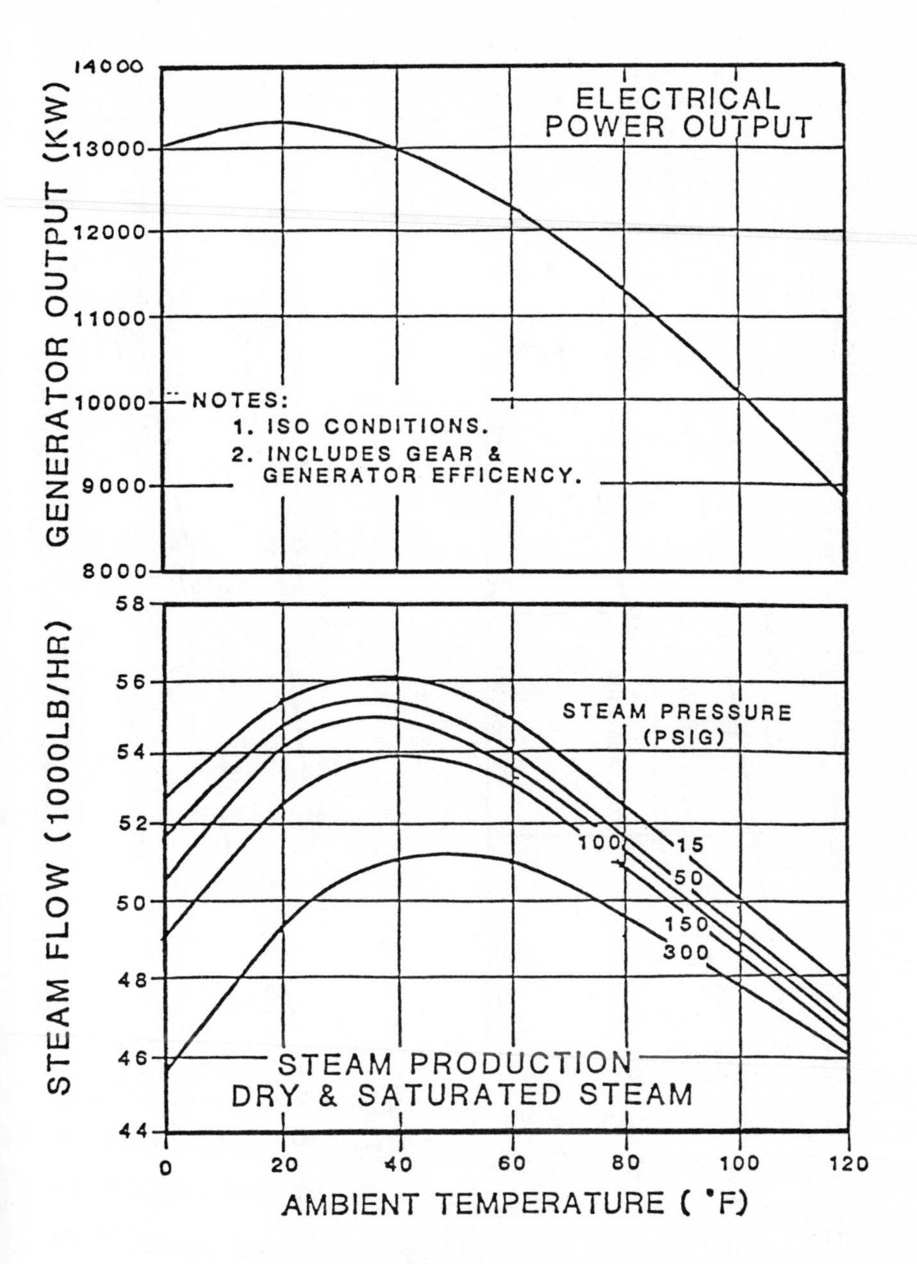

Figure 7-4

STEAM GENERATION CAPABILITY OF A 12 MW GAS TURBINE

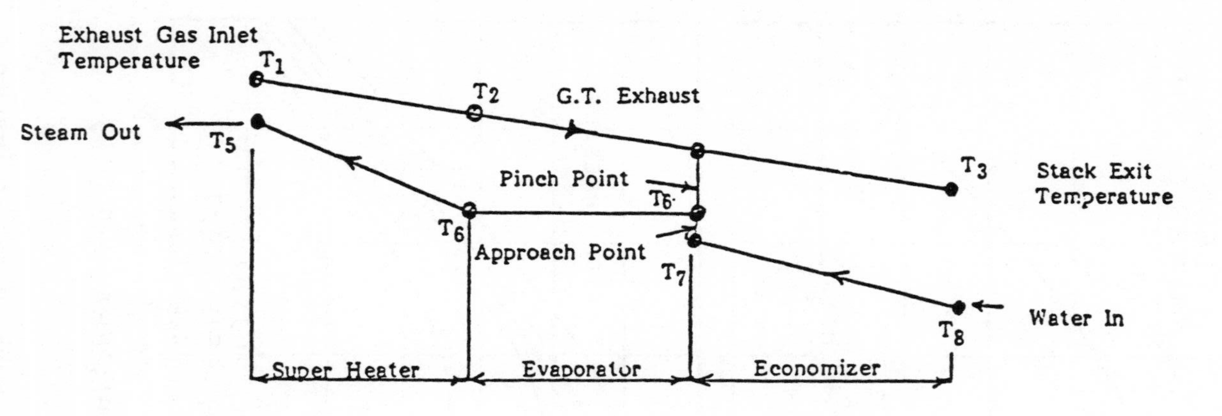

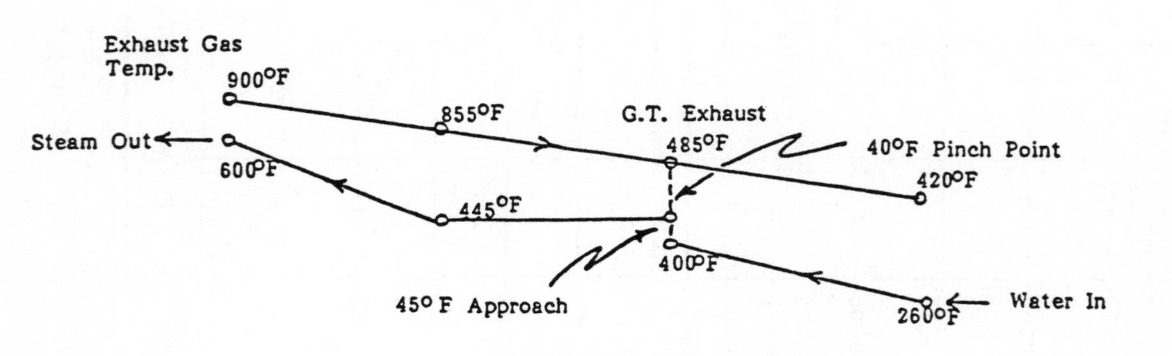

Figure 7-5
TEMPERATURE PROFILES IN A HEAT RECOVERY SYSTEM

h_1 = Enthalpy of Water entering Evaporator

K = Radiation Loss Factor (.985)

It must be noted that any process fluid may be utilized (in the heat recovery unit) such as Dowtherm or Therminol fluids.

Gas Turbine Heat Recovery Units

The waste heat recovery system is a critically important subsystem of a cogeneration system. In the past it was viewed as a separate "add-on" item. This view is being changed with the realization that good performance, both thermodynamically and in terms of reliability, grows out of designing the heat recovery system as an *integral part* of the overall system. Excellent design guidelines are provided in reference 4.

Some important points and observations relating to gas turbine waste heat recovery are:

- *Multipressure Steam Generators* – These are becoming increasingly popular. With a single pressure boiler there is a limit to the heat recovery because exhaust gas temperature cannot be reduced below steam saturation temperature. This problem is avoided by the use of multipressure levels.

- *Off-Design Performance* – This is an important consideration for waste heat recovery boilers. Gas turbine performance is effected by load, ambient conditions and gas turbine health (fouling, etc.), and this can affect exhaust gas temperature and airflow rate. Adequate considerations must be given as to how steam flows (LP & HP) and superheat temperatures vary with changes in gas turbine operation. This is discussed in reference 5.

- *Back Pressure Considerations (Gas Side)* – These are important as excessively high back pressures create performance drops in gas turbines. Very low pressure drops would call for a very large heat exchanger and more expense. Typical pressure drops are 8-10" H_2O.

Gas Turbine Fuels

Most gas turbines in cogeneration application use natural gas as the primary fuel with No. 2 fuel oil as a backup. Gas turbines can, however, burn a wide variety of fuels including crude and residual fuels. Natural gas typically has a Lower Heating Value (LHV) of 950 Btu/Ft^3, and No. 2 fuel oil a LHV of 18,500 Btu/lb. Natural gas has several advantages: (1) it is cleaner, (2) has lower NO_X emissions, and (3) requires no storage. Table 7-1 shows some important fuel characteristics.[10]

If line pressure is low, a fuel gas compressor may be required to boost fuel pressure to above gas turbine compressor discharge pressure to allow injection into the combustors. Reciprocating, Screw or Centrifugal compressors may be used for this purpose.

EMISSION CONSIDERATIONS

With the growing movement for environmental protection, the area of Gas Turbine Emissions has become a critical one for cogeneration projects. EPA regulations and negotiations relating to emissions and offsets by states in an area are best handled by professionals with experience in this area. EPA standards are a function of turbine energy consumption and intended use. Rules and decisions regarding emissions are in a state of flux, which is why expert help in addressing this area is crucial.

The trend towards higher firing temperatures and the use of heavy fuels creates problems with respect to Nitrogen Oxide (NO_X) formation. There are two sources of NO_X: (1) Fuel Bound NO_X, and (2) Thermal NO_X. Fuel bound NO_X is created by the reaction of fuel bound Nitrogen with combustion air (i.e., it is related to the fuel). Thermal NO_X is produced by the oxidation of atmospheric nitrogen at high firing temperatures.

There are several NO_X suppression methods:

- *Steam Injection* – This involves the injection of steam into the combustor to lower combustion temperatures and hence curtail NO_X formation. The output of the turbine is also increased by virtue of the increased mass flow rate. There is also a small increase in thermal efficiency. Steam injection is extensively used.

Fuel Properties

	Kerosene	Diesel Fuel Burner Fuel #2	Oil #2	JP-4	High-Ash Crude Heavy Residual	Typical Libyan Crude	Navy Distillate	Heavy Distillate	Low-Ash Crude
Flash point °F	130/160	118-220	150/200	<RT	175/265		186°F	198	50/200
Pour point °F	−50	−55to+10	−10/30		15/95	68	10°F		15/110
Visc. CS @ 100°F	1.4/2.2	2.48/2.67	2.0/4.0	.79	100/1800	7.3	6.11	6.20	2/100
SSU		34.4					45.9		
Sulfur %	.01/.1	.169/.243	.1/.8	.047	.5/4	.15	1.01	1.075	.1/2.7
API gr.		38.1	35.0	53.2			30.5		
Sp. gr. @ 100°F	.78/.83	85	.82-.88	.7543@60°F	.92/1.05	.84	.874	.8786	.80/.92
Water & ded.			0			.1%wt			
Heating value $\frac{Btu}{lb}$	19300/19700	18330	19000/19600	18700/18820	18300/18900	18250		18239	19000/19400
Hydrogen %	12.8/14.5	12.83	12./13.2	14.75	10/12.5			12.40	12/13.2
Carbon residue									
10% bottoms	.01/.1	.104	.03/.3			2/10			.3/3
Ash ppm	1/5	.001	0/20		100/1000	36ppm			20/200
Na + K ppm	01.5		0/1		1/350	2.2/4.5			0/50
V	0/.1		0/.1		5/400	0/1			0/15
Pb	0/.5		0/1		0/25				
Ca	0/1	0/2	0/2		0/50				

Table 7-1
FUEL PROPERTIES OF SOME TYPICAL FUELS[10]

- *Water Injection* – This involves a principle similar to the above, but can result in increased maintenance problems in the combustor. Also, NO_X and Carbon Monoxide (CO) exist in an inverse manner. An increase in water injection would reduce NO_X but *increase* CO. Typical water injection rates vary from .5 to 1.2 times the fuel flow rates. The water should be very low in dissolved solids and is typically double deionized. Water injection increases work output but decreases the thermal efficiency.
- *Use of Catalytic Converters* – These are capable of reduction of 60-80% NO_X which is brought about by the reaction of NO_X with ammonium hydroxide to produce nitrogen and water in the presence of a catalyst. Selective catalytic reduction (SCR) has recently been declared BACT (Best Available Control Technology) in the state of California.

  There is extensive R&D work underway for "dry combustor design" in which NO_X control is accomplished by attempting to inhibit the creation of NO_X. This is done by partial burning and cooling air quenching within a reconfigured combustor.

CLOSED CYCLE COGENERATION

Closed cycle gas turbines are essentially externally fired gas turbines which allow the use of any type of fuel including water, coal and coal-derived fuels. Closed cycle gas turbines have been used for several years in Europe. Figure 7-6 shows a closed cycle concept as described in reference 6. Cycles such as these provide the following advantages:

- Multi-feed Capability
- Efficient Fuel Utilization
- High Part Load Efficiency
- Suitability for Small Scale Design.

In the closed cycle concept, exhaust gas *must* be cooled (so it can be returned to the compressor inlet). This is done by steam

Figure 7-6
CLOSED CYCLE COGENERATION CYCLE WITH
VARIABLE RECUPERATOR BYPASS

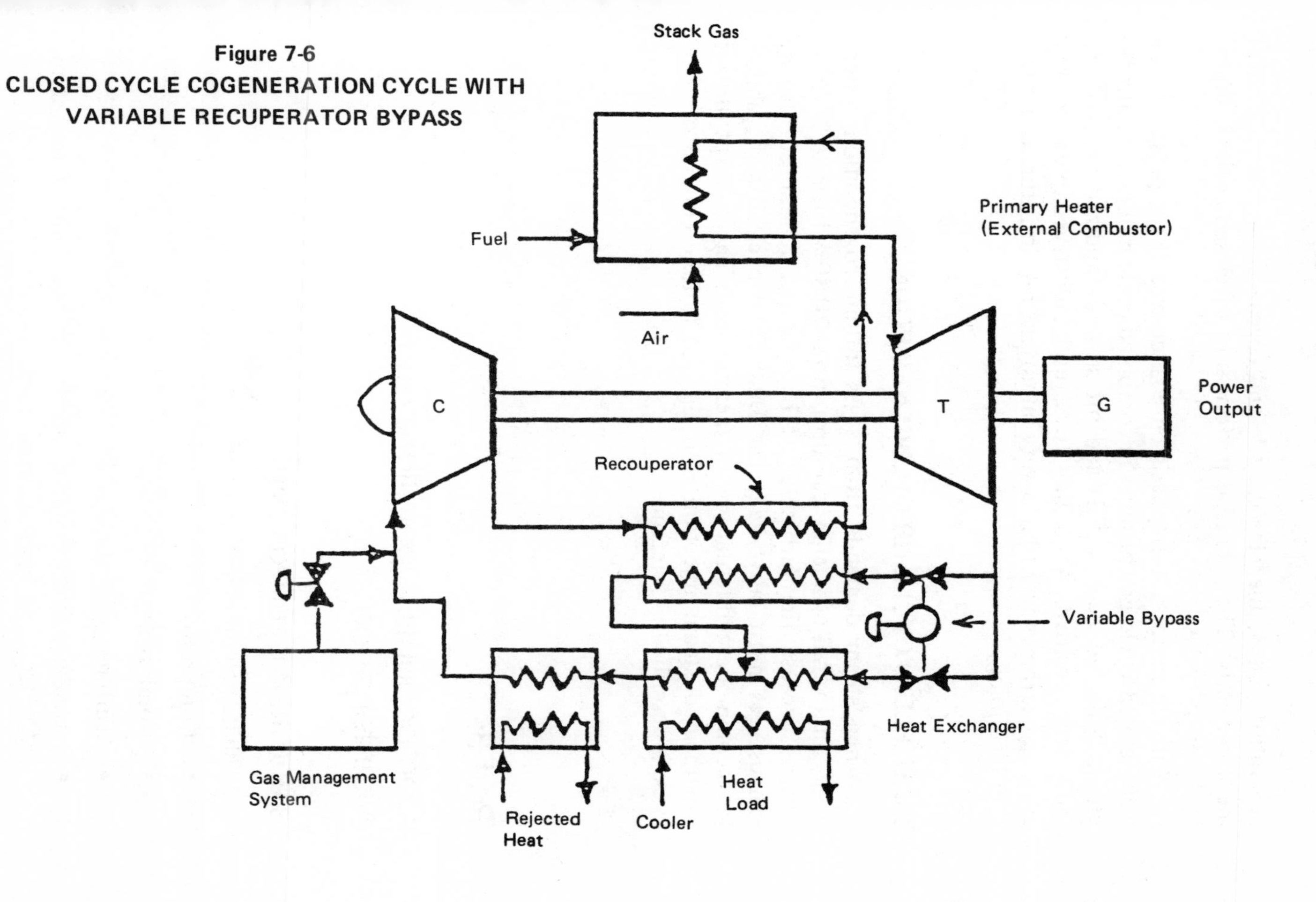

generation. Thus, the closed cycle and cogeneration concepts intermesh perfectly. The part load performance of closed cycles is superior to open cycles.

With an important emphasis on coal-derived fuels, prospects for using closed cycle gas turbines in combined cycle modes with atmospheric fluidized beds (AFB) and pressurized fluidized beds (PRB) look very positive. According to several industry experts, closed cycle gas turbines will play a very important role in future cogeneration technology.

DUAL FLUID CYCLE (DFC) COGENERATION

When steam demands from a gas turbine cogeneration system drop, the overall efficiency can drop from 70-80% down to 20-30%. This is because the turbine must be operated at reduced load or exhaust gas must be vented. The Dual Fluid cycle has the ability to accommodate fluctuations in thermal load. The concept utilizes superheated steam injection into the gas turbine combustor. This cycle is described in reference 7.

Part B:

Components, Operations and Maintenance

GAS TURBINE COMPONENTS

The key components that make up a gas turbine are

- Compressor
- Combustor
- Turbine
- Auxiliary Systems and Controls – Filters, Lube, Seal systems, Controls and Instrumentation.

Compressors

A compressor is a device which pressurizes working fluids (in this case, air). The turbocompressor transfers energy by dynamic means from a rotating member to the continuously flowing air. The two types of compressors used in gas turbines are axial and centrifugal. Some small gas turbines employ a centrifugal compressor or a combination of an axial compressor followed by a centrifugal unit. The larger units employ axial flow compressors.

An axial flow compressor compresses its working fluid by first accelerating the fluid and then diffusing it to obtain a pressure increase. The fluid is accelerated by a row of rotating airfoils or blades (the rotor) and diffused by a row of stationary blades (the stator). The diffusion in the stator converts the velocity increase gained in the rotor to a pressure increase. One rotor and one stator constitute a "stage" in a compressor. A compressor usually consists of several stages. One additional row of fixed blades (inlet guide vanes) is frequently used at the compressor inlet to ensure that air enters the first-stage rotors at the desired angle. In addition to the stators, a diffuser at the exit of the compressor further diffuses the fluid and controls its velocity when entering the combustors.

In an axial compressor, air passes from one stage to the next with each stage raising the pressure slightly. By producing low pressure increases on the order of 1.1:1-1.4:1, very high efficiencies can be obtained. The use of multiple stages permits overall pressure ratios to 18:1 and above.

Figure 7-7 shows the pressure, velocity and total enthalpy variation for flow through several stages of an axial compressor. As indicated in Figure 7-7, the length of the blades, and the annulus area, which is the area between the shaft and shroud, decreases through the length of the compressor. This reduction in flow area compensates for the increase in fluid density as it is compressed, permitting a constant axial velocity.

Figure 7-8 shows what happens to airflow as it enters the compressor. In this figure, V represents the absolute velocity of the flow. The relative velocity (velocity "seen" by the rotating blade) is represented by W. U represents the blade tip speed. $U = \pi DN/720$ where D = inches, N = RPM and U is ft/sec. This diagram represents "one stage." There may be 16 stages in the axial flow compressor. Off-

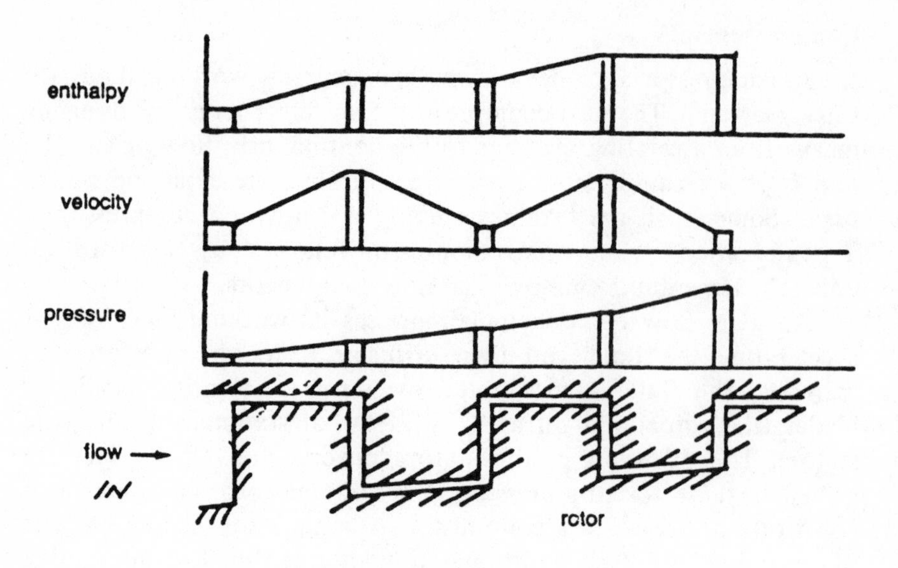

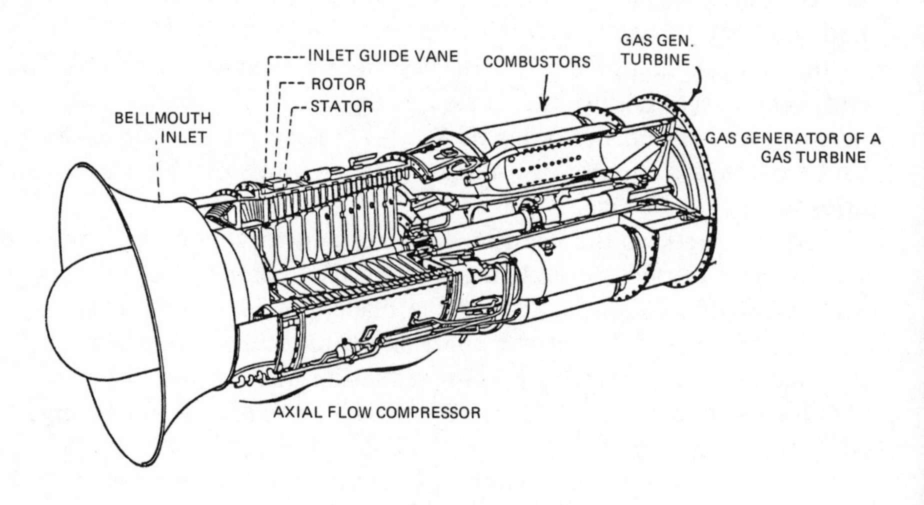

Figure 7-7
AXIAL FLOW COMPRESSOR: p, v & h VARIATIONS, CONSTRUCTION AND PERFORMANCE MAP

Continued ⟶

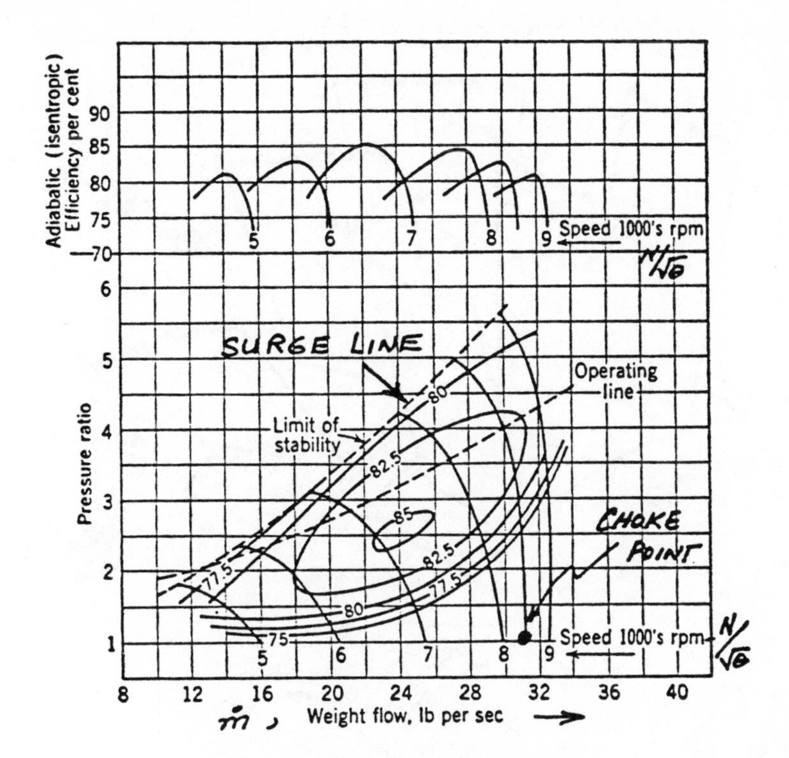

Performance Map for Axial Flow Compressor

design operation (changed N, airflow rate, dirty compressor, etc.) can change the velocity triangles cumulatively and cause certain problems known as stall, surge, rotating stall, etc.[8]

Typically, with compressor pressure ratios in excess of 8:1, variable stator geometry or multiple spool compressors, different operating speeds are required. Variable stators allow "good" velocity triangles to be obtained (i.e., minimized losses).

As was mentioned in an earlier section, 60-70% of total work produced by the turbine is consumed by the compressor; the gas turbine performance can be substantially improved by intercooling the compressor. This, however, increases the complexity of the cycle and is not commonly used. The Japanese Moonlight Project involving an advanced 143 MW gas turbine combined cycle with a pressure ratio of 55:1 utilizes both intercooling and reheat.

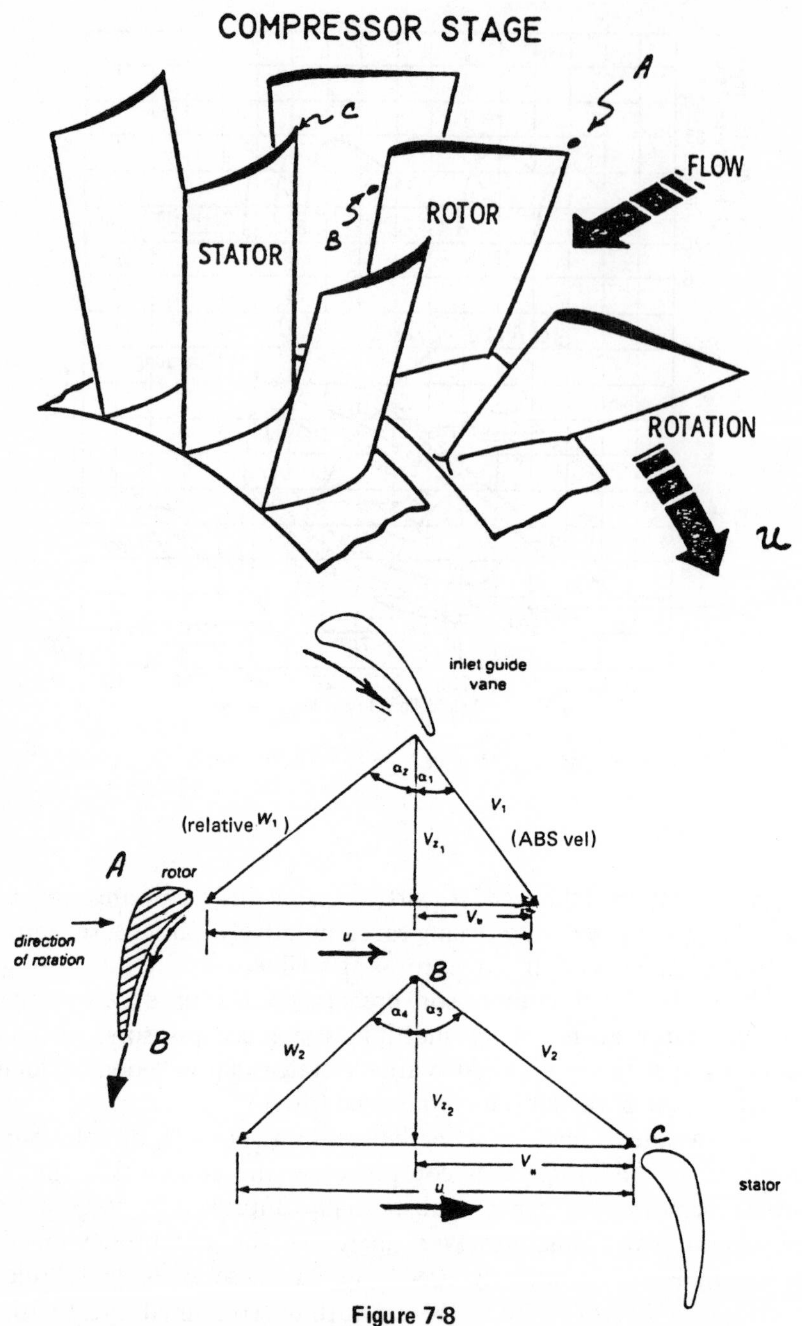

Figure 7-8
VELOCITY TRIANGLES FOR AN AXIAL FLOW COMPRESSOR

Axial flow compressor design is a complex process, but high compressor efficiencies can be obtained. The pressure rise is always acting "against" the flow direction and hence stability problems and flow breakdowns can always occur.

Centrifugal compressors are used on some smaller gas turbine engines. In certain cases, a combination axial and centrifugal compressor may be used. Pressure ratios per stage can vary from 2:1 to as high as 12:1 on some experimental wheels. The centrifugal compressor is less efficient than an axial flow compressor, but it is more stable (larger surge-choke margin). Figure 7-9 shows a typical centrifugal compressor section.

Combustors

All gas turbine combustors perform the same function; they increase the temperature of the high-pressure gas. There are different methods of combustor arrangement on a gas turbine:

- Tubular (side combustors)
- Can-annular
- Annular.

Tubular (side combustors) – These designs are found on large industrial turbines, especially European designs, and some small vehicular gas turbines. They offer the advantages of simplicity of design, ease of maintenance and long life due to low heat release rates. These combustors may be of the "straight-through" or "reverse-flow" design. In the reverse-flow design, air enters the annulus between the combustor can and its housing, usually a hot-gas pipe to the turbine. Reverse-flow designs have minimal length.

Can-annular and Annular – In aircraft where frontal area is important, either can-annular or annular designs are used to produce favorable radial and circumferential profiles because of the great number of fuel nozzles employed. The annular design is especially popular in new aircraft designs; however, the can-annular design is still used because of the developmental difficulties associated with annular designs. Annular combustor popularity

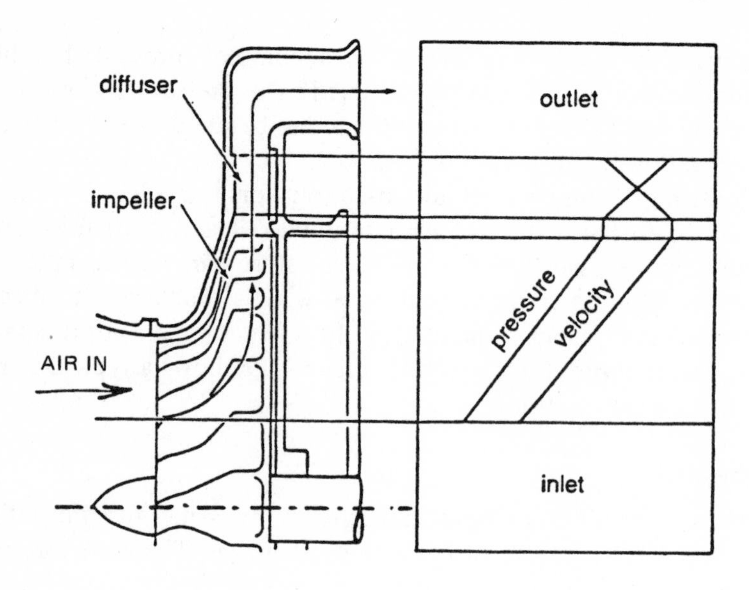

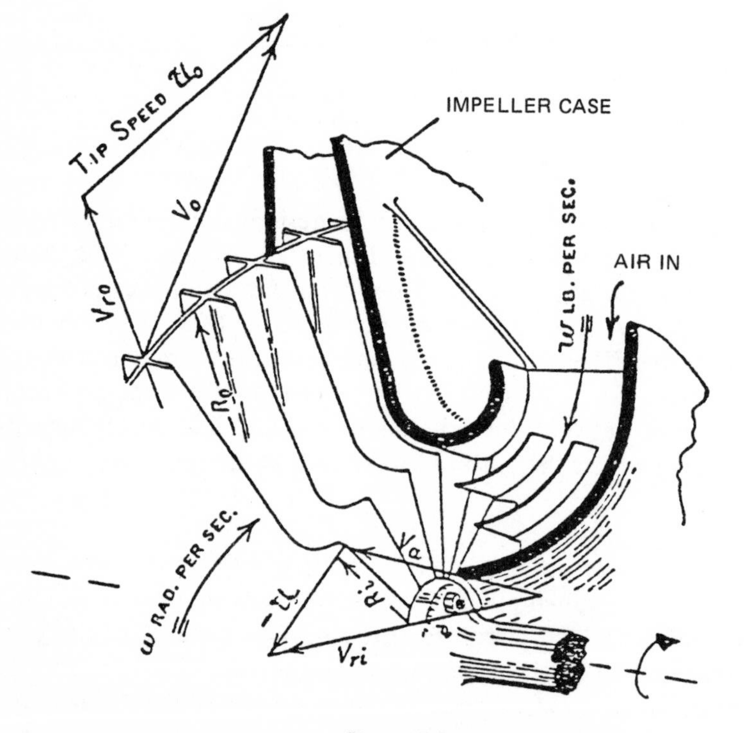

Figure 7-9
CENTRIFUGAL COMPRESSOR DETAILS

increases with higher temperatures or low-Btu gases, since the amount of cooling air required is much less than in can-annular designs due to a much smaller surface area. Can-annular combustors can be of the straight-through or reverse-flow design. If can-annular cans are used in aircraft, the straight-through design is used, while a reverse flow design may be used on industrial engines. Annular combustors are almost always straight-through flow designs.

Despite the many design differences, all gas turbine combustion chambers have three features: (1) a recirculation zone, (2) a burning zone (with a recirculation zone which extends to the dilution region), and (3) a dilution zone. The function of the recirculation zone is to evaporate, partly burn, and prepare the fuel for rapid combustion within the remainder of the burning zone. Ideally, at the end of the burning zone, all fuel should be burnt so that the function of the dilution zone is solely to mix the hot gas with the dilution air. The mixture leaving the chamber should have a temperature and velocity distribution acceptable to the nozzles and turbine metallurgy.

The pressure drop in a combustor can vary from 2-8%. This results in increased fuel consumption. Typical combustor efficiencies are 95-98%. A large amount of excess air is utilized for cooling purposes (400% Stoichiometric). This is the reason why ample oxygen exists in the exhaust enabling supplemental firing in the waste heat recovery unit. Figure 7-10 shows a can-annular combustor showing the different combustion zones.

Turbines

There are two types of turbines used in gas turbines. These consist of the axial flow type and the radial inflow type. The axial flow turbine is used in most gas turbines.

Axial-Flow Turbines – The axial-flow turbine, like its counterpart the axial-flow compressor, has flow which enters and leaves in the axial direction. There are two types of axial turbines: (1) impulse, and (2) reaction. The impulse turbine has its entire enthalpy drop in the nozzle, therefore producing a very high velocity entering the rotor. The reaction turbine divides the enthalpy drop in the nozzle and the rotor. Figure 7-11 is a

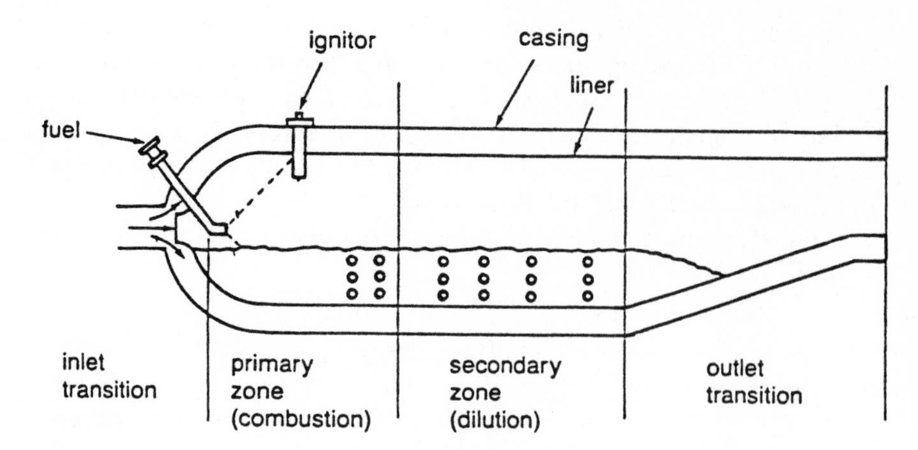

Figure 7-10
CAN ANNULAR COMBUSTOR WITH ZONES

schematic of an axial flow turbine, along with some other information.

Radial Inflow Turbine – The radial inward flow radial turbine has been in use for many years. Basically a centrifugal compressor with reversed flow and opposite rotation, the inward flow radial turbine is used for smaller loads and over a smaller operational range than the axial turbine. Radial inflow turbines are used on several hot gas expanders. Axial turbines have enjoyed tremendous interest due to their low frontal area, making them suited to the aircraft industry. However, the axial machine is much longer than the radial machine, making it unsuited to certain applications. Radial turbines are used in turbochargers, in some types of expanders, and on a few gas turbines in the low power range.

GAS TURBINE OPERATIONS AND MAINTENANCE CONSIDERATIONS

Some important factors that impact gas turbine operating reliability are fuel considerations (including fuel quality, filtration, treatment), compressor air filtration, compressor washing and health monitoring.

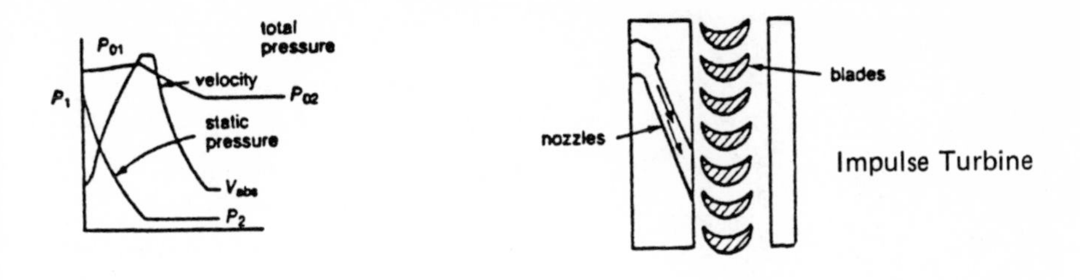

Impulse Turbine

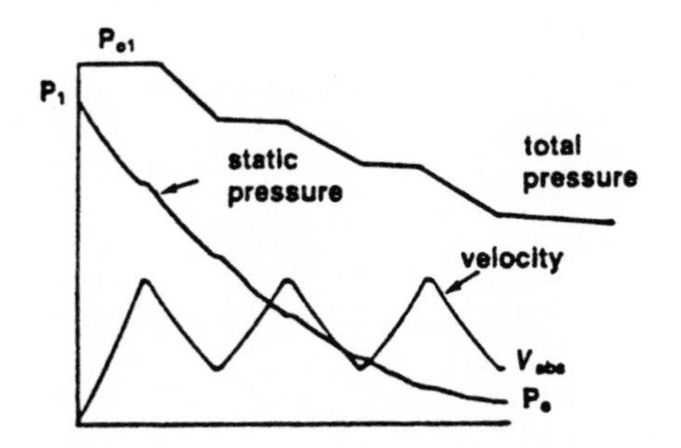

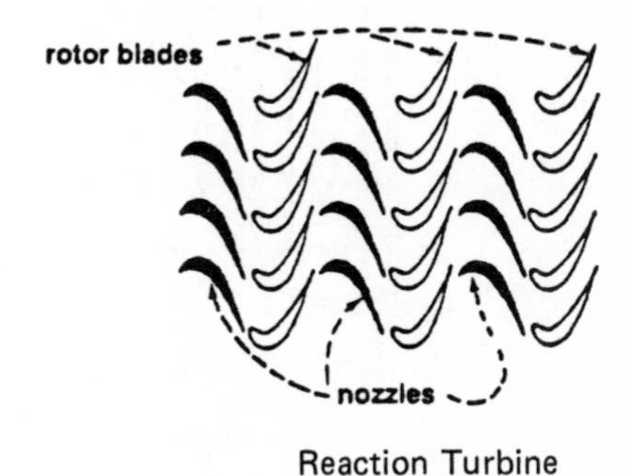

Reaction Turbine

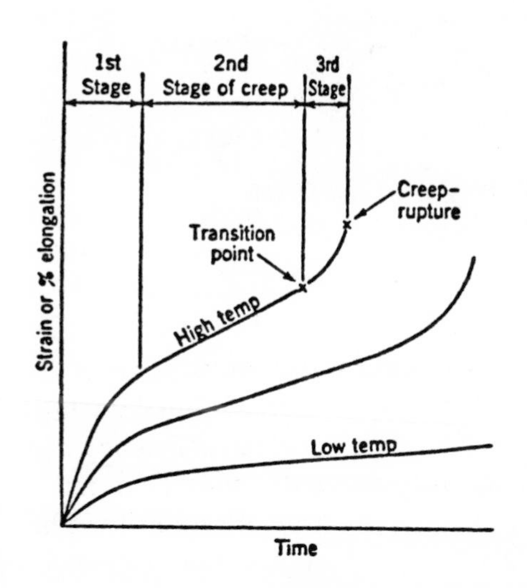

CREEP PHENOMENA

- Blades are at High Temp. & Stress Levels (2 ounce blade can exert centrifugal force of 2 tons)
- Acceptable blade & nozzle metal temperature are obtained by bleed air cooling from compressor.

Figure 7-11
AXIAL FLOW TURBINE DETAILS

Continued ⟶

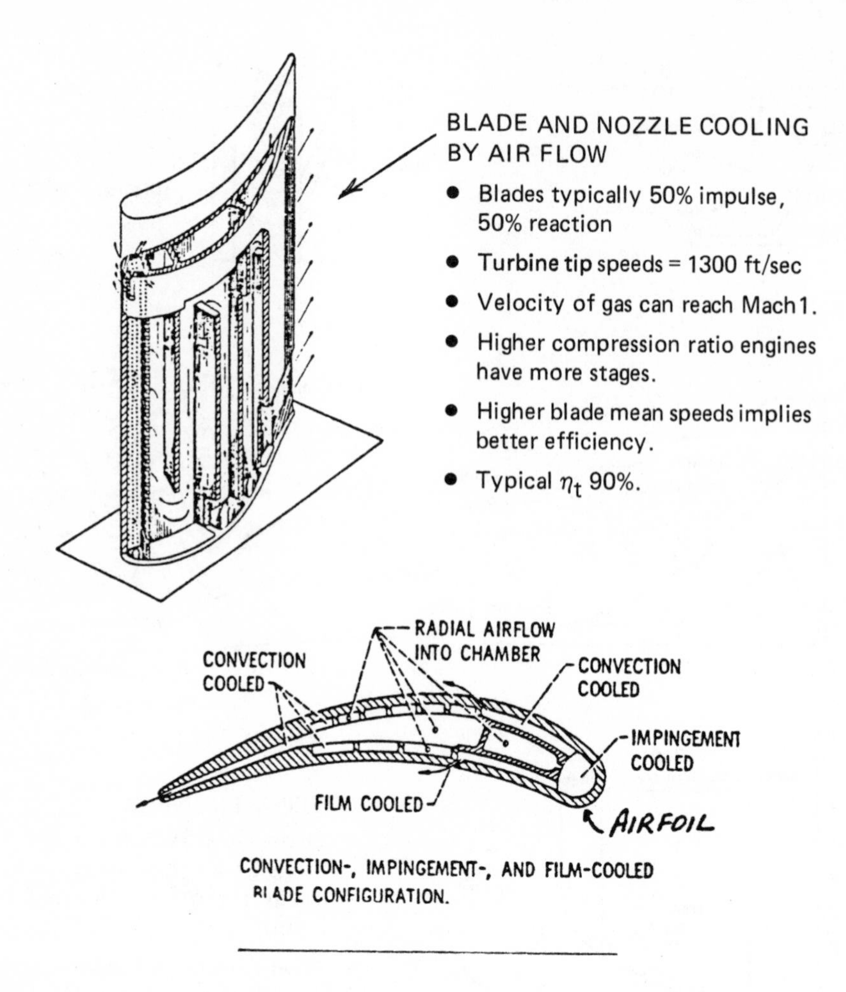

CONVECTION-, IMPINGEMENT-, AND FILM-COOLED BLADE CONFIGURATION.

It is important to note that maintaining the gas turbine at its peak operating efficiency is most important from a fuel savings standpoint. In cogeneration applications, good overall efficiency can be obtained even with the use of turbines with poor heat rates, because rejected heat is utilized for steam generation or for the process, but the use of low efficiency gas turbines should be carefully evaluated.

Typically, compressor fouling is responsible for a drop in load output caused by a drop in pressure ratio, air flow rate, compressor efficiency, and overall thermal efficiency. Of these, the most evident is the drop in load—10 to 20% not being uncommon with a dirty compressor. Because of this, effective and well scheduled compressor washing procedures are of paramount importance. Compressor degradation (fouling) can be responsible for 70-80% of the output loss. On a simple cycle turbine, a 5% drop in airflow and compressor efficiency can cause an output drop of around 13% and a heat rate increase of about 5%.

Compressor Air Filtration and Compressor Cleaning

In past filter reviews it has been found that an air filter that controls corrodents also does a good job on foulants. Based on this experience, this discussion will be more weighted toward corrosion control since the foulant control is a healthy byproduct.

Modern-day gas turbines operate at a turbine inlet temperature of about 1700°F to 2100°F which makes them highly sensitive to corrosion problems. This calls for care to be taken with respect to air filtration and compressor washing. Even with good filtration, salt can collect in the compressor section. During the collection process of both salt and other foulants, an equilibrium condition is reached quickly, after which reingestion of large particles occurs. This reingestion has to be prevented by the removal of salt from the compressor prior to saturation. The rate at which saturation occurs is highly dependent on filter quality. In general, salts can safely pass through the turbine when gas and metal temperatures are less than 1000°F. Aggressive attack will occur if the temperatures are much higher. During cleaning, the actual instantaneous rates of salt passage are very high, together with greatly increased particle size.

A good air filtration system should include:

- *First Stage Filter* – This should be an inertial filter which should remove all particles greater than 10 microns.
- *Second Stage Filter* – This should be a roughing filter capable of removing particles to the order of 5 microns.
- *Third Stage Filter* – This should be a high efficiency filter with filter material that is hydrophobic (unwettable) in nature. In this recommended scheme, the first and second filtration stages are to ensure a long life and efficient operation of the third-stage, high-efficiency filter as well as to keep operating costs low.

Even with good air filtration, salt deposits will occur in the compressor. As the air moves through the compressor toward the combustion section, it is heated and compressed causing removal of the remaining moisture from the airborne salt particles. These particles are deposited heavily in the first few stages sometimes going back as far as half way through the compressor. In general, the condensation nuclei that pass through the compressor without being entrapped, can also pass through the turbine without depositing themselves, or adhering to the hot parts. A difficulty arises however when the salt that has been collected within the compressor stages becomes so thick that large flakes are reingested into the engine. When this occurs, the local concentration of salt in the air immediately surrounding this large flake is extremely high. These salt flakes actually have sufficient mass to stick firmly to the turbine hot parts, and are responsible for many gas turbines suffering from hot corrosion damage. Saturation of the compressor with salts thus still remains a problem and must be dealt with. Compressor cleaning will continue to be necessary even with a machine utilizing an efficient air filter, but on a greatly reduced schedule. Compressor cleaning is very important for the efficient and trouble-free running of a machine and must be carried out on a regular basis.

With an adequate high efficiency air filtration system, cleaning of the compressor should not be needed more than once in four months. Careful performance calculations can indicate when the compressor is beginning to foul and cleaning can then become a part of a scheduled maintenance program rather than something done on an emergency or haphazard basis. Another philosophy that has met with considerable success is a time based wash, e.g., once every two weeks. The time wash period is a function of environmental conditions.

As salt is soluble in water, one method of cleaning is to wash out the salt using a soap-and-water solution. The soap is added to breakdown or remove any oil which may have been condensed within the compressor. While steam cleaning of the compressor is very efficient, water washing, using a water soap mixture, is the most efficient method of cleaning. This cleaning is most effective when carried out in several steps which involve the application of a soap-and-water solution, followed by several rinse cycles. Each rinse cycle involves the acceleration of the machine to approximately 50% of the starting speed, after which the machine is allowed to coast to a stop. A soaking period follows during which the soap solution may work on dissolving the salt. If the machine is cleaned under full operating conditions using rice, wheat, walnut shells or some other solid material injected into the inlet, the foulants and corrosive elements deposited within the compressor are removed rapidly and flushed through the turbine.

If the machine is running at substantial load and the gas temperatures are in the region where the salt can be fused, it will have the opportunity of combining with the sulphur (which is present in the fuel), and forms sodium sulphate. The sodium sulphate may then fuse, become sticky and deposit itself on the hot parts of the turbine causing severe and rapid attack which is characteristic of hot corrosion (sulphidation). If the metal temperatures and gas temperatures are less than 1000°F, then the salt will pass safely through the turbine as it never has the opportunity to become molten.

The method recommended for determining whether or not the foulants have a substantial salt base is to soap wash the turbine and collect the water from all drainage ports available. Dissolved salts in the water can then be analyzed. The amount of water passed through the machine and the time at which it was collected allows an establishment of proper washing procedure. A program of simple performance tests should be conducted to detect compressor fouling. Pressure drops over filter elements should be recorded daily. Water washing a machine under power must be extremely closely controlled to prevent the possibility of liquid water impinging upon hot turbine parts. A carefully controlled water flow using turbine operating parameters as control functions can produce efficient cleaning with minimal or no damage to the machine and with a very short period of reduced power output.

Good filtration can have a significant impact on reducing fouling, reducing compressor erosion, cooling air hole blockage and a host of other problems. Money should never be "saved" by compromising a filtration system.

Fuel System Considerations

The construction of the fuel system and of the actual oil or fuel used by a gas turbine must be investigated and considered from a reliability operating standpoint during the design phase of the project.

It is commonly known that the movement of combustible fuels by barge is always accompanied by the ingress of some water into the fuel. One gallon of sea water in 1,000,000 gallons of oil is sufficient to render that fuel unfit for combustion in gas turbines without the use of a fuel cleanup system. The presence of sodium (both airborne and within the fuel) and Vanadium (within the fuel) has a very corrosive effect on hot section life. Most turbine manufacturers insist on levels at or less than 1ppm.

Water will generally settle and carry salt out of the oil during storage periods. If pure water is added to the fuel, it will nearly always be more salty when recovered. This means that the water concentration near the bottom of the storage tank will always be higher than the water concentration near the upper surface of the liquid. It is important that a floating suction system be used for delivering gas turbine fuel. Floating suctions should be installed in these tanks, which would preclude the pickup of moisture settled at the bottom of the tank.

Depending on the water binding characteristics of the particular fuel involved, several methods of filtration are available. Paper filters will under no condition render sufficient fuel cleanliness to preclude the ingress of water into the turbine. These filters have been found to remove water, provided the water is not bound in emulsion within the fuel. If the water is bound within emulsion, it is necessary to inject a demulsifier and break the emulsion either by centrifuge or by an electrostatic precipitation.

Electrostatic precipitation involves the addition of between 5-10% of pure water within a mixing system. The water and salt dissolved by this added water are then removed either by centrifuge or by electrostatic precipitation. Heating for viscosity reduction may be required.

Adequate fuel filtration and salt reduction can be achieved by these methods. Edge type filters such as the Cuno type and paper filters do not generally provide adequate protection against entrained water in fuel. The presence of Vanadium in the liquid fuel is usually inhibited by the addition of Magnesium Sulphate (Epsom Salts). An excessive amount of this material causes turbine fouling and consequently more frequent turbine washing.

Fuel nozzles, which draw their fuel from a common manifold, are always more sensitive to plugging and poor distribution of their fuel spray pattern. Careful monitoring and control of the temperature spread in the exhaust gas must be maintained. Exhaust gas temperature spreads of more than 60°F or sudden changes of 30°F above normal variation in a carefully monitored machine should be grounds for concern. If high exhaust gas temperature spreads are detected, the machine should be shut down and the burner nozzles cleaned or inspected to ensure that uneven distribution of fuel does not occur. If this spread is not corrected very rapidly, combustor distress will occur.

In the case of a gas turbine fired by natural gas, the problem of Sodium and Vanadium is not present. However, care should be taken that liquid slugs are not present in the fuel. This can be achieved by adequate fuel preheating, the use of knock-out drum and scrubbing. Excessive heat densities applied during the preheating can, however, cause problems with coking of condensibles.

Coatings

The superalloys currently used by gas turbine engine manufacturers for fabrication of blades and vanes for turbine hot sections are predominantly either nickel or cobalt-based with compositions emphasizing high temperature mechanical properties. Their chemical formulations, therefore, preclude sufficient amounts of those elements, primarily cromium and aluminum, that could make these alloys inherently resistant to the chemical reactions encountered during engine service. Though the operating environment of turbine blades and vanes varies widely depending on engine design and application, the requirements for protective coatings are basically the same, namely, resistance to hot corrosion and oxidation for long intervals of cyclical engine operation.

A wide variety of high temperature coatings for new and engine run turbine components are available. These differ substantially in complexity and cost, thereby enabling the user to obtain the desired protection for his parts at reasonable expense.

Currently, several types of compressor blade coatings are available which may (1) improve erosion resistance, (2) improve corrosion resistance, and (3) provide aerodynamically smooth blades and consequently better compressor efficiency.

For gas turbine engines, some important features that aid maintenance include the availability of borescope ports, vibration and condition monitoring systems, diagnostics for electrical and control circuits, availability of special tools and facilities for maintenance alignment, balancing, etc., accessibility of auxiliary pumps, compressor washing systems and the ability to conduct quick filter changes.

Decisions made during the design phase can have a significant impact over the cogeneration project life cycle. Key items to be reviewed are:

- Ease of Access of High Maintenance Items – Combustion liners, fuel nozzels, first-stage nozzles, etc. For example, some gas turbine units have combustion liners and caps with hinges enabling them to be unbolted and swung to one side without the use of special lifting tackle.
- Ease of Access for Inspection – Borescopes, inspection doors should be provided to gain access to the compressor intake and hot section. In the case of large gas turbines and heat recovery units, adequate safety handrails, walkways, and ladders should be provided. Provisions should be made for the availability of lighting, power points for tools, etc.
- Auxiliary Systems and Controls for the Gas Turbine must be accorded special attention – Arrangements should be made for Duplex filters so that one filter can be cleaned while the other is in service. Lube oil systems should have provisions for tie-ins for centrifuges so that oil cleaning can be accomplished as fast as possible.

Experience has shown that in the case of gas turbines, a major area of deficiency has been the accessories including clutches, pumps, couplings, air compressor dampers and instrumentation and controls. The user must take extra care when specifying in this area. Instrumentation and dampers become more unreliable with higher environmental temperatures. Reference 9 provides some guidelines for reliability improvement.

In order to keep the gas turbine operating reliably, a number of inspections are required. Inspection could range from daily "walk-around" checks to major inspections which call for gas turbine disassembly. Daily inspections should include the following checks:

- Lube Oil Levels
- Oil Leaks Around Engine
- Loose Piping Fasteners
- Visual Inspection of Inlet Filter and Filter Differential Pressure.

The interval between detailed inspections depends on the machine type, manufacturer's recommendations and environmental conditions.

The use of borescopes for inspecting gas turbine engines allows the following benefits:

- Internal Onsite Checks Without Disassembly
- Allows Accurate Planning and Scheduling of Maintenance Actions
- Monitors Condition of Internal Components
- Increased Ability to Predict Required Parts, Special Tools and Skilled Manpower.

Turbine inspection and overhaul guidelines are provided by the manufacturer and should include effects of number of starts, fuel being used, and usage. Manufacturers usually have equations that provide combustion inspection interval, hot gas path inspection and major inspection as a function of:

- Starting Frequency
- Fuel Type
- No. of Fired Hours at Peak Load
- No. of Fired Hours at Base Load
- No. of Starts/Hour
- Fuel Type.

Table 7-2 provides some guidelines for operation and maintenance life of industrial gas turbines.[10]

Table 7-2. OPERATION AND MAINTENANCE LIFE OF AN INDUSTRIAL GAS TURBINE[10]

Type of Application Fuel	Type Inspection (Hrs. of Operation)				Expected Life (Replacement) Hrs. of Operation		
	Starts/hr	Service	Minor	Major	Comb. Liners	1st Stage Nozzle	1st Stage Buckets
BASE	*	+	+	+	+	+	+
Nat. gas	1/1000	4500	9000	28,000	30,000	60,000	100,000
Nat. gas	1/10	2500	4000	13,000	7500	42,000	72,000
Distillate oil	1/1000	3500	7000	22,000	22,000	45,000	72,000
Distillate oil	1/10	1500	3000	10,000	6000	35,000	48,000
Residual	1/1000	2000	4000	5000	3500	20,000	28,000
Residual	1/10	650	1650	2300			
SYSTEM PEAKING ×							
Nat. gas	1/10	3000	5000	13,000	7500	34,000	
Nat. gas	1/5	1000	3000	10,000	3800	28,000	
Distillate	1/10	800	2000	8000			
Distillate	1/5	400	1000	7000			
TURBINE PEAKING ×							
Nat. gas	1/5	800	4000	12,000	2000	12,000	
Nat. gas	1/1	200	1000	3000	400	9000	
Distillate	1/5	300	2000	6000			
Distillate	1/1	100	800	2000			

* 1/5 = One start per five operating hours
× No residual usage due to low load factor and high capital cost
Base = Normal maximum continuous load
System Peaking = Normal maximum load of short duration and daily starts.
Turbine Peaking = Extra load resulting from operating temperature 50-100°F above base temperature for short durations.
Service = Inspection of combustion parts, required downtime approximately 24 hours
Minor = Inspection of combustion plus turbine parts, required downtime approximately 80 hours
Major = Complete inspection and overhaul, required downtime approximately 160 hours
Note: Maintenance times are arbitrary and depend on manpower availability and training, spare parts and equipment availability, and planning. Boroscope techniques can help reduce downtime.

Combustor Inspections: These involve a short shutdown where the combustor is dismantled, cleaned and inspected. The fuel nozzles, liners, transition pieces are carefully checked for cracks and other problems. A borescope inspection of the first-stage nozzle and blades is also done.

Hot Gas Path Inspection: This requires disassembly of the turbine casing. All nozzles and blades should be checked for cracks, foreign object damage, hot corrosion or fouling.

Major Inspections: This involves a complete removal of the gas turbine case and a check of all components–including axial flow compressor, bearings, seals and hot section.

Condition monitoring is becoming very popular and by monitoring certain flows, temperatures, pressures and vibration and applying computational techniques along with trending, a good indication of machine health is arrived at. Condition monitoring can often allow the pinpointing of the component undergoing mechanical or aerodynamic distress.[11]

With aeroderivative gas turbines the whole gas generator can be removed and a spare installed. This ensures a high availability. A 25MW aeroderivative gas generator can be removed and replaced in 10-12 hours with a crew of 5 people.

CONCLUSION

The gas turbine engine will play an increasingly important role in the future cogeneration market. Future developmental efforts will further this prime mover's attractiveness and popularity. Careful attention during system specification and design can ensure a high availability from gas turbine based cogeneration systems.

NOMENCLATURE

T = Temperature
P = Pressure
h = Enthalpy
s = Entropy

$\dot{m}$	=	Mass Flow Rate
C_p	=	Specific Heat at Constant Pressure
Q	=	Heat Consumption
LHV	=	Lower Heating Value
PR	=	Pressure Ratio
D	=	Diameter
U	=	Tip Speed
V	=	Absolute Velocity
W	=	Relative Velocity

Subscripts

t	=	Turbine
c	=	Compressor
b	=	Combustor
a	=	Air
f	=	Fuel
1,2..	=	Cycle State Points
th	=	Thermal
o	=	Output

Greek

γ	=	Rate of Specific Heats = C_p/C_v
η	=	Efficiency

REFERENCES

[1]M.P. Boyce, *et al.*, "Optimization of Various Gas Turbine Cycles," *Proceedings of the Third Turbomachinery Symposium,* Texas A&M University, October 1974.

[2]R.W. Foster-Pegg, "Calculations and Performance Adjustments for On-Site Conditions," *Sawyer's Turbomachinery Handbook,* 2nd Edition, Vol. 2.

[3]C.B. Meher-Homji and G. Mani, "Combustion Gas Turbine Power Enhancement by Refrigeration of Inlet Air," *Proceedings of the Fifth Annual Industrial Energy Conservation Technology Conference,* April 18-21, 1983. (This covers a review of techniques–evaporative cooling, intercooled cycles, expansion cooling, compression and absorption cycles, 12 references are included.)

[4]V. Ganapathy, *Applied Heat Transfer,* Penwell Books, 1982. (This is an excellent detailed text covering practical design aspects. A large number of easy to use nomograms are provided for quick analysis. Treats the area of combined cycles and cogeneration in detail.)

[5]A. Pasha and R.W. Precious, "Gas Turbine Heat Recovery–A Complex Design Process," *Power,* October, 1982, p. 89.

[6]W.M. Crim, W.E. Fraize and G.A. Malone, "Closed Cycle Cogeneration for the Future," ASME Paper #84-GT-272.

[7]L. Kosla *et al.,* "Inject Steam in a Gas Turbine but not just for NO_X Control," *Power,* February 1983.

[8]C.B. Meher-Homji and A.B. Focke, "Performance and Vibration Monitoring for the Prevention of Gas Turbine Airfoil Failures," ASME Publication H-331, 6th Biennial, ASME Failure Prevention & Reliability Conference, Cincinnati, September 10-13, 1985.

[9]C.B. Meher-Homji and A.B. Focke, "Reliability, Availability and Maintainability Considerations for Gas Turbine Cogeneration Systems," Sixth Annual IECT Conference, April 1984.

[10]M.P. Boyce, *Gas Turbine Engineering Handbook,* Gulf Publishing, 1982.

[11]M.P. Boyce, C.B. Meher-Homji and G. Mani, "The Development and Implementation of Advanced Online Monitoring and Diagnostic Systems for Gas Turbines," 1983 Tokyo International Gas Turbine Conference, Paper IGTC-94.

CHAPTER 8

Coal-Fired Fluidized Bed Combustion Cogeneration

Cabot Thunem, P.E. & Norm Smith, P.E.

INTRODUCTION

Energy managers are increasingly selecting fluidized bed combustion as a reliable and economical technology for production of steam. This trend is most noticeable in the larger sizes, i.e., above 100,000 lb/hr steam production.

Literally hundreds of small fluidized bed combustors have been installed for incineration of coal, municipal sludge, and other low-grade forms of biomass wastes. Acceptance of the technology as a prime source of reliable steam in industrial applications has generally been at ratings of 150 psi steam pressure and below.

In addition to selection of fluidized bed combustion due to environmental and multifuel advantages, the concept coupled with cogeneration has gained relatively wide acceptance. While there are still relatively few fluidized bed combustion cogeneration systems installed and operating, there are numerous installations in the planning and construction stage.

The following outlines a general approach to fluidized bed combustion in the 40,000 lb/hr size. Three different systems currently in the planning stage are discussed. Various factors such as steam requirements, electric usage, load profiles, and rate structures are considered.

Fluidized bed combustion (FBC) permits use of numerous fuels such as coal wash tailings or wood waste and they are usually priced below premium fuels, while maintaining compliance with environmental regulations. The following sections illustrate practical applications of FBC technology for providing steam and electricity (cogeneration) simultaneously. Cogeneration provides an extension of the technology while enhancing the economic feasibility of the solid fuel combustion.

Although stoker firing of coal is a well established technology, FBC was selected as the design approach. The rationale for selection of fluidized bed combustion is twofold, namely:

1. *Emission Control* – Even though small boilers do not have to meet current emission standards, the trend is toward more stringent controls. Fluidized bed combustion provides sulfur and nitrogen oxides emission control in lieu of installing scrubbers or other treatments.
2. *Multifuel Capability* – Use of fluidized bed combustion provides wide latitude in the switching and blending of fuels.

COAL-FIRED PROCESS

The base case of FBC process with coal firing illustrates a fluidized bed combustor generating steam at a pressure of 650 psig, 750°F. The steam is reduced from the boiler pressure to the distribution pressure in a steam turbine with shaft energy converted to electricity. The major items of equipment required include the fluidized bed combustor, baghouse, deaerator, turbine generator, boiler feed pumps, water treatment, and storage silos for coal, waste, and limestone. The turbine generator can be a noncondensing or condensing type. For the condensing operation, a condenser, cooling towers, and cooling water treatment system are also required. Typical operating conditions for three case studies are shown in Table 8-1. Figure 8-1 illustrates the FBC components, including the baghouse.

GENERATION/CONDENSING SYSTEM DESCRIPTION

Two coal-fired cogeneration methods are investigated—a noncondensing turbine (case 1) and a condensing turbine (case 2). Peak net electric generation is 1,044 kW for the noncondensing design. With the condensing turbine, peak generation is 3,600 kW. A third system of no cogeneration (case 3) with 150 psig steam and no turbine is also examined.

Table 8-1
PROCESS SPECIFICATIONS

	Noncondensing(1) Cogeneration	Condensing(2) Cogeneration	No(3) Cogeneration
Feedwater Temperature	220°F	220°F	220°F
Steam Conditions	650 psig, 750°F	650 psig, 750°F	150 psig, sat.
Steam Flow	40,000 lbs/hr (pph)	40,000 pph	40,000 pph
Flue Gas Temperature	300-350°F	300-350°F	300-350°F
Ca/S(molar) Ratio	1-2.5	1-2.5	1-2.5
Excess Air	15%	15%	15%
Combustion Temperature	1,550°F	1,550°F	1,550°F
Extraction Pressure	150 psig	150 psig	—
Condensing Pressure	—	3 in. Hg abs.	—

Source: Stanley Consultants

Notes: (1) Case 1 – Industrial Plant
(2) Case 2 – Industrial Plant
(3) Case 3 – Industrial Plant

The modular designed components selected for this system are shown on Figure 8-2. Approximately an 80′ x 80′ building would be required for installation of all component modules. Although a smaller structure is feasible where site restrictions prevail, we recommend a larger facility for ease of service.

Each of these modules can be factory-assembled, prewired, and tested prior to delivery. Shop assembly will not only reduce construction times and cost, but should also minimize start-up problems.

The boiler system selected is also shop-assembled. It is an atmospheric fluidized bed coal-firing unit. The unit as purchased includes all fans, controls, motor control centers, feed hoppers, and waste removal components.

The turbine-generator skid includes controls and breaker connections in addition to the turbine gearbox and generator. The condensing turbine operates on 3 in. Hg backpressure with a 150 psig extraction port.

The condenser system is sized for condensation of the full 40,000 lb/hr steam generation. Pumping capacity is installed with 100%

backup. Cooling towers are also shipped as modules. However, the condensing system overall will require substantial site mechanical and electrical work.

The cooling water treatment is a two-stage unit sized for 15% makeup. Pumps, cooling controls, and piping are all included in the package. Biocide and antiscaling treatment are included.

A demineralizer system is installed for treatment of the boiler feedwater makeup. The makeup system is sized with 100% backup during regeneration. Other required equipment includes bulk storage tanks and chemical feed pumps for acid and caustic regeneration and a regeneration waste neutralization tank. This equipment will all be provided as a skid-mounted, prewired, factory tested, and assembled unit.

The air supply includes compressors, motors, and surge capacity. The skid-mounted units supply instrument and transport air for the solids conveying systems.

The silo system includes two coal storage silos, a limestone silo, and a spent bed and ash silo. Silos are sized for 10 days' storage.

The dust collector is a modular baghouse. Each component is shop assembled and shipped to the site. The modules are interconnected in the field.

Condensing System Capital Cost – The projected cost for the AFB system with generation/condensing is $7.1 million (Table 8-2). This is a budgetary cost estimate which includes fees and contingency.

GENERATION/NONCONDENSING SYSTEM DESCRIPTION

The overall systems' requirements for noncondensing operation are similar to the condensing system. However, several subsystems are eliminated in a noncondensing design. Items not included are the turbine condensing section, cooling towers, and cooling water treatment, as well as several pumps and miscellaneous piping. The noncondensing turbine operates with a back pressure of 150 psig and generates approximately 0.026 kW/lb steam.

Noncondensing Capital Cost – The estimated cost for the noncondensing system is $5.2 million (see Table 8-2). This cost estimate includes contingencies and design fee.

Table 8-2
SUMMARY CONCEPTUAL COST ESTIMATE

	Noncondensing(1) Cogeneration	Condensing(2) Cogeneration	No(3) Cogeneration
Building	$ 510,000	$ 510,000	$ 70,000
Major Equipment	1,750,000	2,985,000	1,000,000
Ash Handling	190,000	190,000	190,000
Coal Conveying	625,000	625,000	625,000
Limestone Handling	100,000	100,000	100,000
Miscellaneous Equipment	409,000	720,000	50,000
Piping	315,000	315,000	80,000
Electrical Power	480,000	480,000	30,000
Subtotal	4,379,000	5,925,000	2,145,000
Contingencies & Fee	867,000	1,187,000	429,000
Total	$5,255,000	$7,112,000	$2,574,000

Source: Stanley Consultants

Notes: (1) Case 1 - Industrial Plant
(2) Case 2 - Industrial Plant
(3) Case 3 - Industrial Plant

FBC WITHOUT COGENERATION

In addition to the condensing and noncondensing systems, a budgetary estimate is included for an FBC system with no cogeneration. This system has major capital cost savings in that typically existing water treatment and boiler feed are adequate, no electrical interconnections for parallelling with the utility need to be included, and the building can be considerably smaller than for the systems with water treatment and turbine installation. The cost is approximately $2.6 million (Table 8-2).

ANCILLARY POWER REQUIREMENTS

Power requirements for the FBC coal-fired facility with a condensing turbine are approximately 900 hp. This includes all fan, pump, and material-handling motors. For the noncondensing system about 320 hp associated with the cooling tower and condensers is eliminated for approximately 580 hp requirement. For modeling purposes, parasitic losses are 200 kW for the noncogeneration systems and for the noncondensing system. Parasitic losses are 430 kW for the condensing system. The apparent discrepancy between reduced installed horsepower and parasitic losses results from redundant capacity. For instance, cooling tower pumps are sized at 100 hp. Two pumps are installed–one operating and one spare. Since only one operates, connected horsepower is reduced by 200 hp, but the operating kW is reduced by less than 75.

UTILITY INTERCONNECTS

Provision for connecting the generator to the substation is necessary with cogeneration. All transformers for connection to the plant power bus and the substation are included. The generator can operate at 4,160 volts or the substation voltage.

The generator should be equipped with metering, instrumentation, and controls for synchronization and parallel operation with the electric utility. The generator should be capable of supplying power to the plant system as well as back into the local utility's system. Relay and safety requirements vary dramatically among utilities and will affect site specific design cost.

The generator will require a complement of relays for protection of the unit in the event of electrical and mechanical malfunction. Provision will also be made for supply connections from the distribution system to supply the auxiliaries electric power required for emergency boiler operation.

SITE COMPATIBILITY

The fluidized bed combustor, steam-turbine generator, and ancillary equipment are all housed in a new building. The baghouse and stack, along with coal, limestone, and ash storage silos, are exterior to the new building. Steam lines and condensate return will be required for connection of the new system to the existing facility.

ENVIRONMENTAL REGULATIONS

Several guidelines and regulations have been promulgated on the federal and state levels to mitigate environmental concerns resulting from solid fuel combustion. The primary environmental concerns include air emissions, water intake and discharge, and solid waste disposal. These may require environmental and impact assessment.

Air Emissions – The major permitting requirements for air emissions are contained in the federal regulation for Prevention of Significant Deterioration (PSD). When determining the necessity of a PSD permit, two criteria apply:

1. Potential emissions of the new source.
2. Actual increased emissions from the modified source.

The first criterion for PSD applicability governs any increase in emissions from the proposed new source or modification. After consideration of control, if emissions exceed the significant levels listed in Table 8-3, a PSD permit is required. Preliminary estimates of air emissions for all cases utilizing high sulfur coal have been made for sulfur dioxide (174 tons per year), particulate (7 tons per year), and nitrogen oxides (146 tons per year) for a 40,000 lb/hr boiler. These preliminary estimates indicate that sulfur dioxide and nitrogen oxides exceed significant levels. PSD regulations require the following actions:

- A modeling demonstration to show that the increase in emissions of criteria pollutants will not exceed national ambient air quality standards.
- A monitoring program on air quality analysis to determine baseline air quality.
- The application of best available control technology for control of the pollutants subject to this regulation.

All or part of these actions may be required depending on available data or previous air quality monitoring. An applicability determination must be performed and submitted to the United States Environmental Protection Agency to determine if PSD regulations could apply to this project.

In addition to PSD permitting, the proposed new source may be subject to regulations governing nonattainment areas. If an area is

Table 8-3
SIGNIFICANT EMISSION RATES[1]
FOR PSD APPLICABILITY

Pollutant	Emission Rates (tons/yr)
Carbon monoxide	100
Nitrogen oxides	40
Sulfur dioxide	40
Particulate matter	25
Ozone (VOC)	40 (of VOCs)
Lead	0.6
Asbestos	0.007
Beryllium	0.0004
Mercury	0.1
Vinyl chloride	1
Fluorides	3
Sulfuric acid mist	7
Hydrogen sulfide (H_2S)	10
Total reduced sulfur (including H_2S)	10
Reduced sulfur compounds (including H_2S)	10

Source: 40 Code of Federal Regulations 52.21 (b) (23) of the United States
Notes: (1) For all Cases.

classified as not attaining the air quality standards rules governing nonattainment areas, offsets and lowest achievable emission rates may apply. The expected emission rates for the boiler will satisfy best available control technology, but may not be the lowest achievable emission rate applied in nonattainment areas.

New sources must also obtain a state air quality permit. Often, information provided in the PSD permit is sufficient for a state permit. If PSD permitting is not required, the state may request air quality dispersion modeling in addition to the information needed to obtain a permit. The state agency should be consulted before any applications for air permits are made.

Solid Waste Disposal – The solid discharge from a fluidized bed combustor is presently classified as a nonhazardous waste similar to fly ash. These solids may be disposed of in a landfill after obtaining the appropriate permits. This report assumes the waste will be disposed of in the coal mine or in the municipal landfill.

Water Intakes and Discharge – It is not anticipated that a retrofitted facility will have a significant impact on either the water requirements or the water discharges; however, an analysis must be conducted to determine water quality impacts. The actual water inputs and outputs will have to be estimated as part of the permitting process. A new facility must comply with all state and federal water quality requirements for water intake and discharge. In both cases an NPDES permit will probably be required.

Environmental Assessment – As part of the procedure for obtaining permits, an environmental assessment of the project is often required. This assessment summarizes all environmental impacts of the proposed new source and the surrounding environment. This assessment is used to determine if an environmental impact statement should be performed.

CASE STUDY ECONOMIC ANALYSIS/MODEL

An essential ingredient to the success of the FBC technology is its economic feasibility in a variety of applications. Therefore, this chapter includes economic analysis and examples from three different reports conducted by Stanley Consultants.

Projection of the economic feasibility of an FBC cogeneration unit requires development of a comprehensive simulation model. This model enables the manipulation of engineering criteria, fuel costs, and facility operating patterns to determine the conditions under which both switching to solid fuel and installation of cogeneration is economically advantageous.

These case studies were selected to illustrate the effects of local rate structures in usage patterns on selection of three different options: cogeneration with back pressure turbine, cogeneration with a condensing extraction cycle, and FBC with cogeneration. A noncogeneration example is included because many times the incremental payback for on-site power generation is not economically feasible due to rate structure or load usage patterns.

Each case study includes actual base case electric power and steam requirements from an analysis performed by Stanley Consultants. Without the cogeneration project, the electric requirements are met through purchase from a utility company and the steam requirements are met through existing natural gas boilers. Average monthly loads for steam and electric power are inputs into this simulation. Input for electric power costs include both energy and demand components. Electric loads included base load, which is constant for each month, and added summer air conditioning, when appropriate.

Case Study 1, Noncondensing Cogeneration

The first case study is an industrial facility in Des Moines, Iowa. The fuel selected for analysis is unwashed Iowa coal. Both washed and unwashed coal were tested in a 2.25 sq ft fluidized bed combustion system at the University of North Dakota Energy Research Center. Test results indicated that 90% reduction in emissive sulfur control along with oxides of nitrogen emissions between .4 and .5 lbs per million Btu could be obtained with a fuel cost per million Btus fired of $1.33. This $1.33 includes a delivered cost of coal, a delivered cost of limestone, and the total costs for bed material and baghouse ash disposal in a landfill.

While both a condensing and noncondensing system were analyzed for this particular site a simple back-pressure turbine was selected. The estimates of system payback were very similar for both systems. However, the noncondensing system was recommended because the system has significantly reduced complexity, and provides a better match of production/demand.

The nonleveraged simple payback is estimated at 4.3 years. A detailed cash flow analysis for the first 10 years of the project is included in Figure 8-3.

Case Study 2, Condensing Cogeneration

This case study is another small industry which presently pays approximately 4.4¢/kWh for electricity and $4.75 per million Btu for natural gas. Within this utility service area, buybacks are available on a levelized basis of 6.5¢/kWh, which exceeds the purchase cost of electricity. Gas turbines, an FBC coal-fired system with condensing cycle, and an FBC coal-fired system with a back-pressure turbine were analyzed.

The fluidized bed combustion system with steam condensing capability provided the best overall payback. This FBC combustor system also includes the capability to burn wood waste or coal washery waste.

The estimated simple payback for the project is approximately 3.1 years. The accompanying Figure 8-4 illustrates the estimated cash flow components for the first 10 years of operation.

Case Study 3, No Cogeneration

This case study is for an industrial manufacturing facility. The average purchase cost of electricity is approximately 5.3¢/kWh. The sell backs range between 1.4¢ and 2.5¢/kWh, ranging between off-peak winter and on-peak summer. This particular application has a demand of a little over 2,000 MW with an average load factor of 60%. Because of this rather poor load factor, the maximum value displaced electricity is approximately 2.6¢/kWh. The buybacks preclude use of a condensing extraction cycle and the required back pressure severely limits the available electric power from a back-pressure turbine.

Detailed analysis of the maximum value of the displaced electricity indicated that approximately $100,000.00 of electric purchase could be displaced at an incremental capital cost of over $2 million dollars. As a result of this rather poor payback on the electric portion, we recommended that the coal-fired system be considered without cogeneration. The major benefit derived for this plant is a displacement of natural gas purchase by coal.

The estimated simple payback for this system is 2.8 years. The detailed cash flow analysis is presented in Figure 8-5.

CONCLUSIONS

Results of analyses for three separate projects indicate that coal-fired fluidized bed combustion is technically, environmentally, and economically feasible for a variety of applications. Cogeneration, utilizing the steam produced, usually enhances the project. However, each application must be analyzed separately to determine the effects of the interactions between steam and electric load, the interaction between the cogenerator and the electric grid, the existing rate structure and interconnection agreements, and fuel availabilities.

Although not discussed in specific detail, fluidized bed combustion offers a choice of a wide variety of nonconventional fuels in addition to coal. These fuels typically are coal washery tailings from coal mines and wood waste. Dozens of other fuels have been burned successfully in fluidized bed combustion. Depending on the cost and availability of such fuels, FBC can be shown to be even more attractive than conventional fuels and truly the combustion technology of the future.

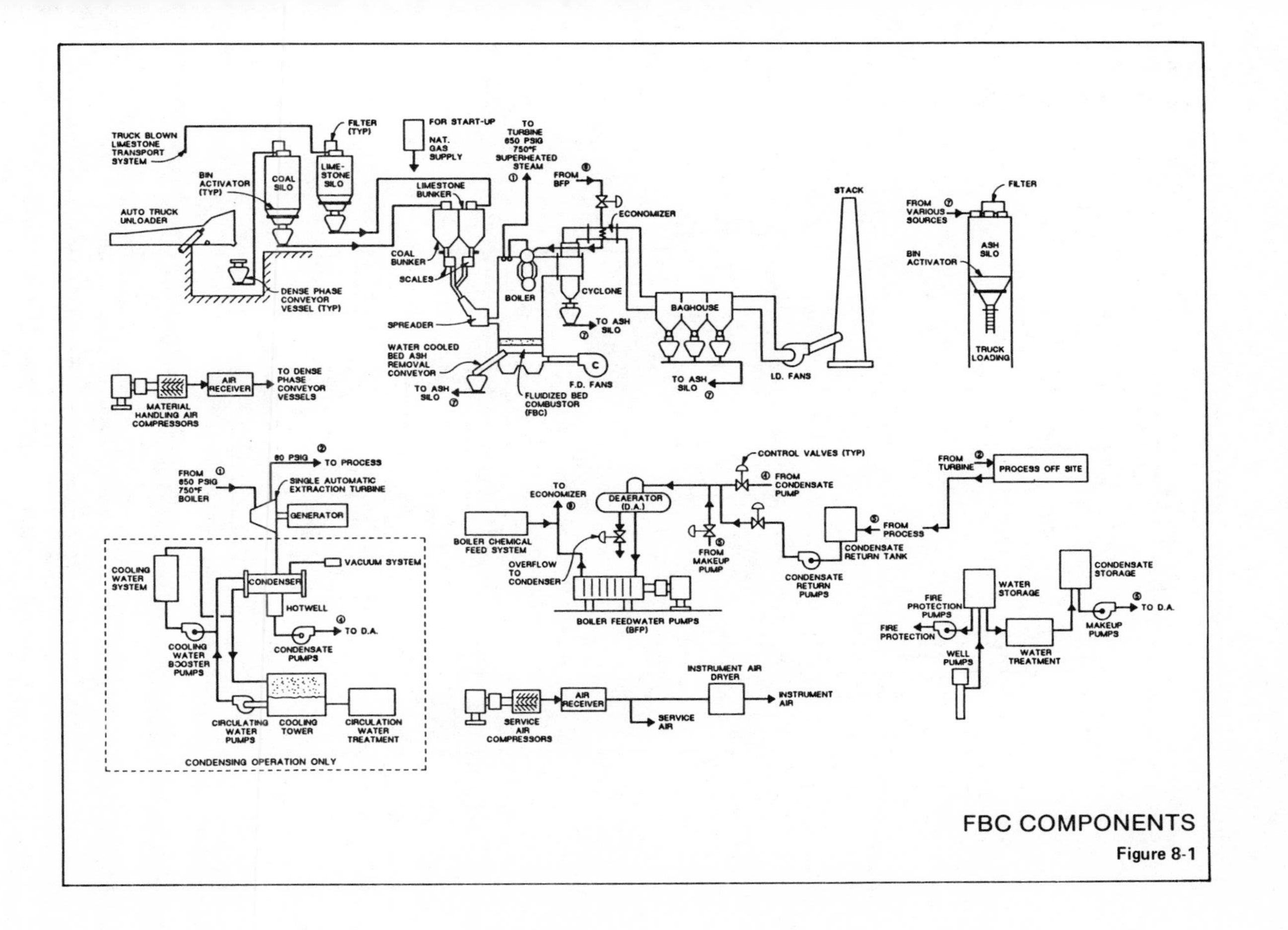

FBC COMPONENTS
Figure 8-1

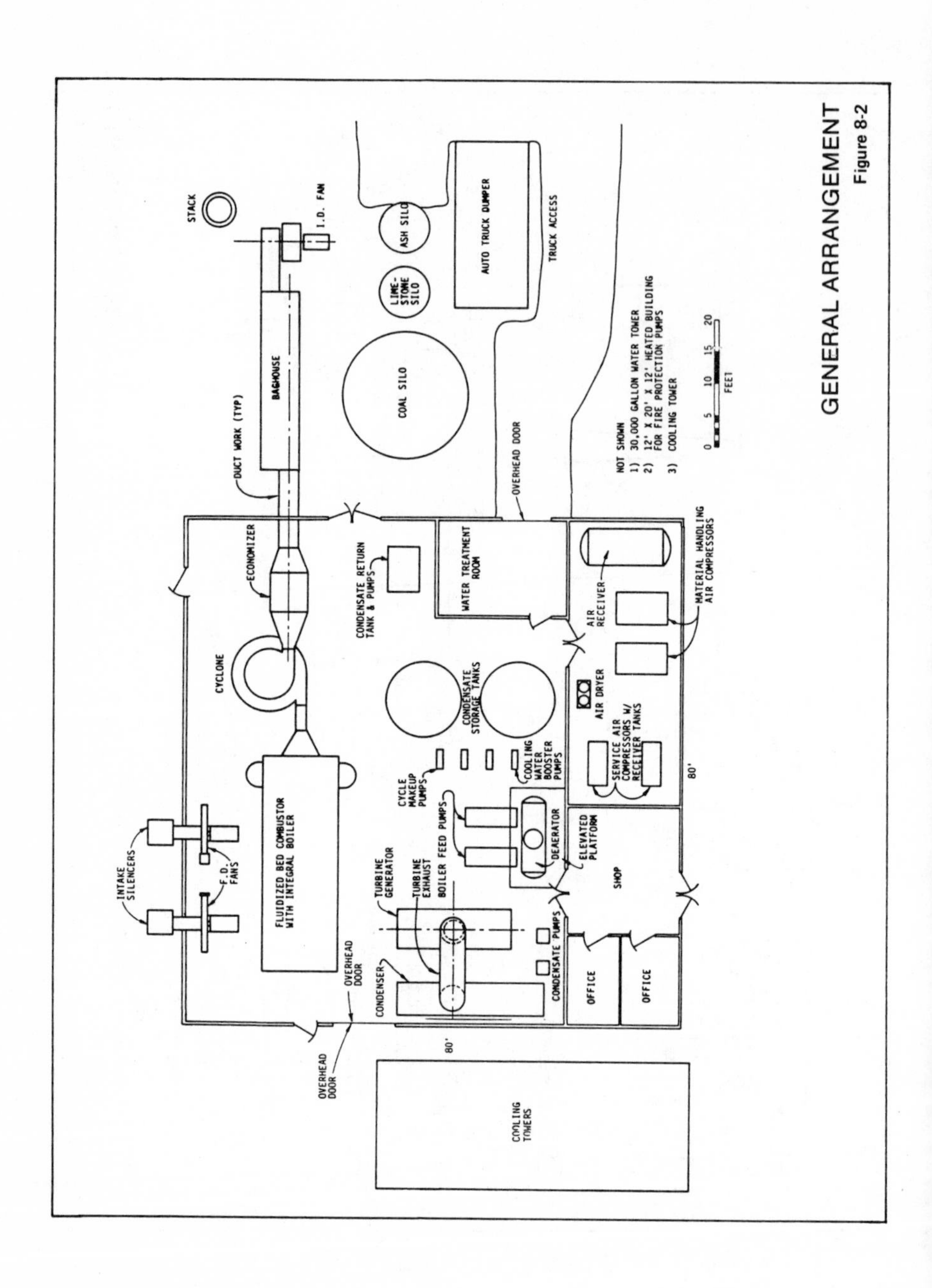

GENERAL ARRANGEMENT
Figure 8-2

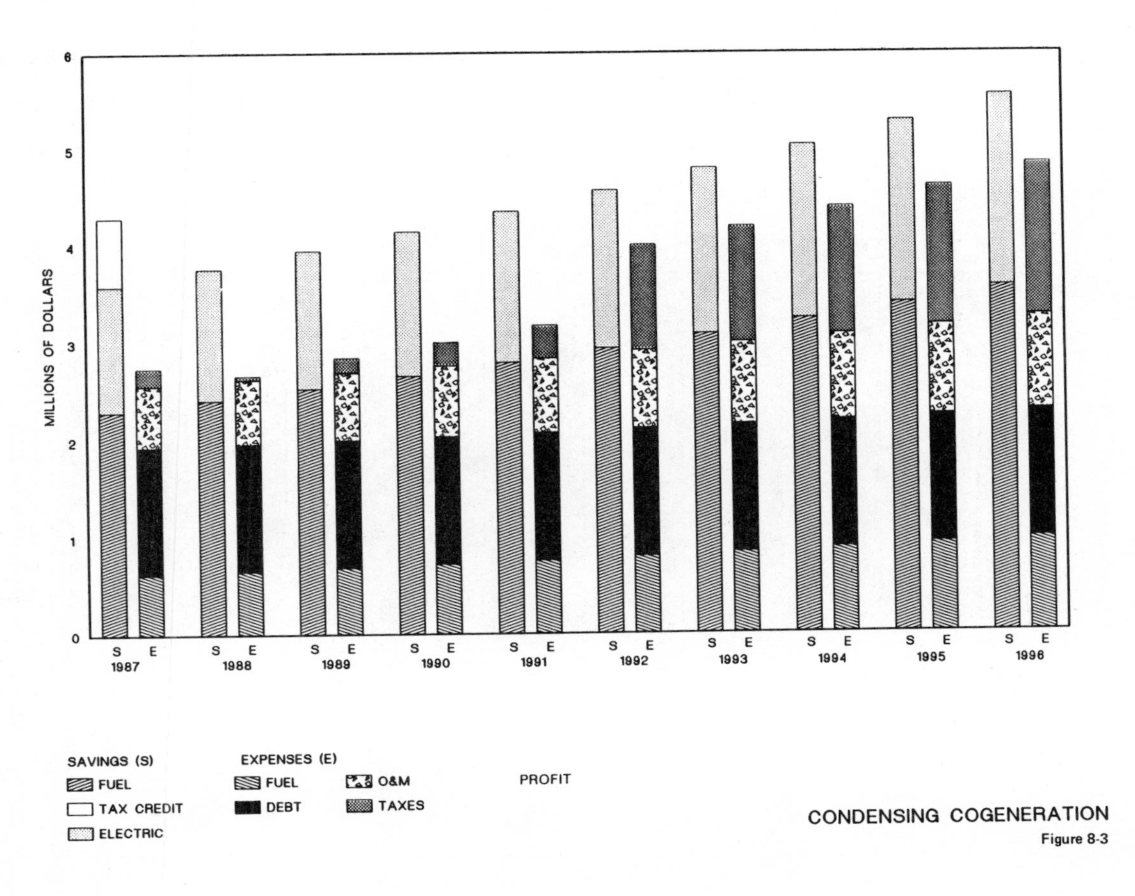

CONDENSING COGENERATION
Figure 8-3

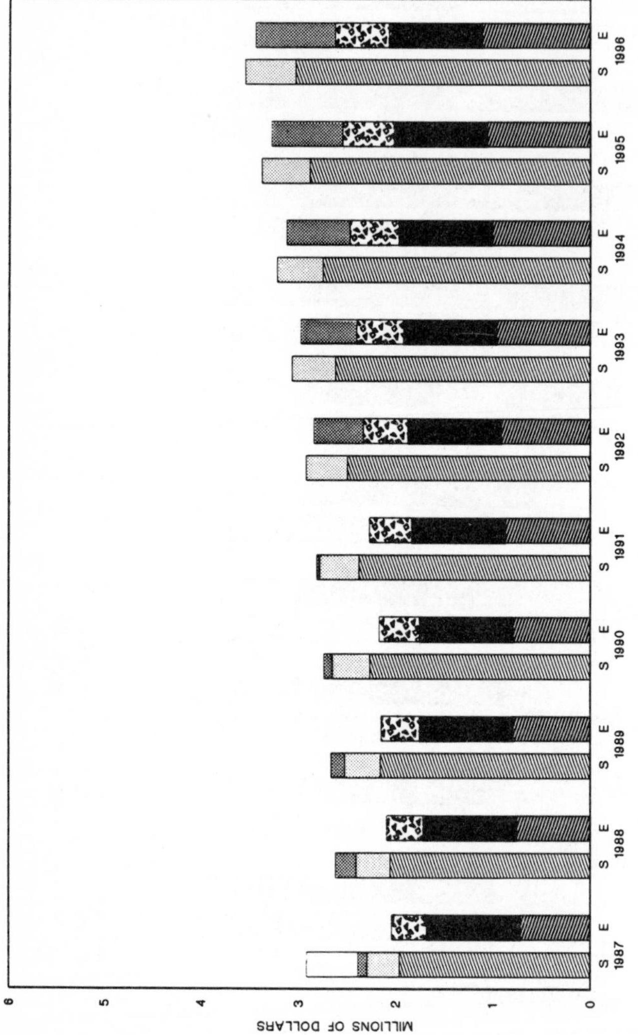

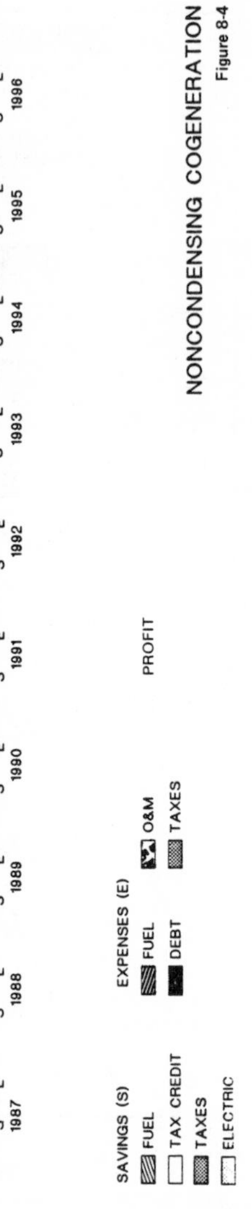

NONCONDENSING COGENERATION
Figure 8-4

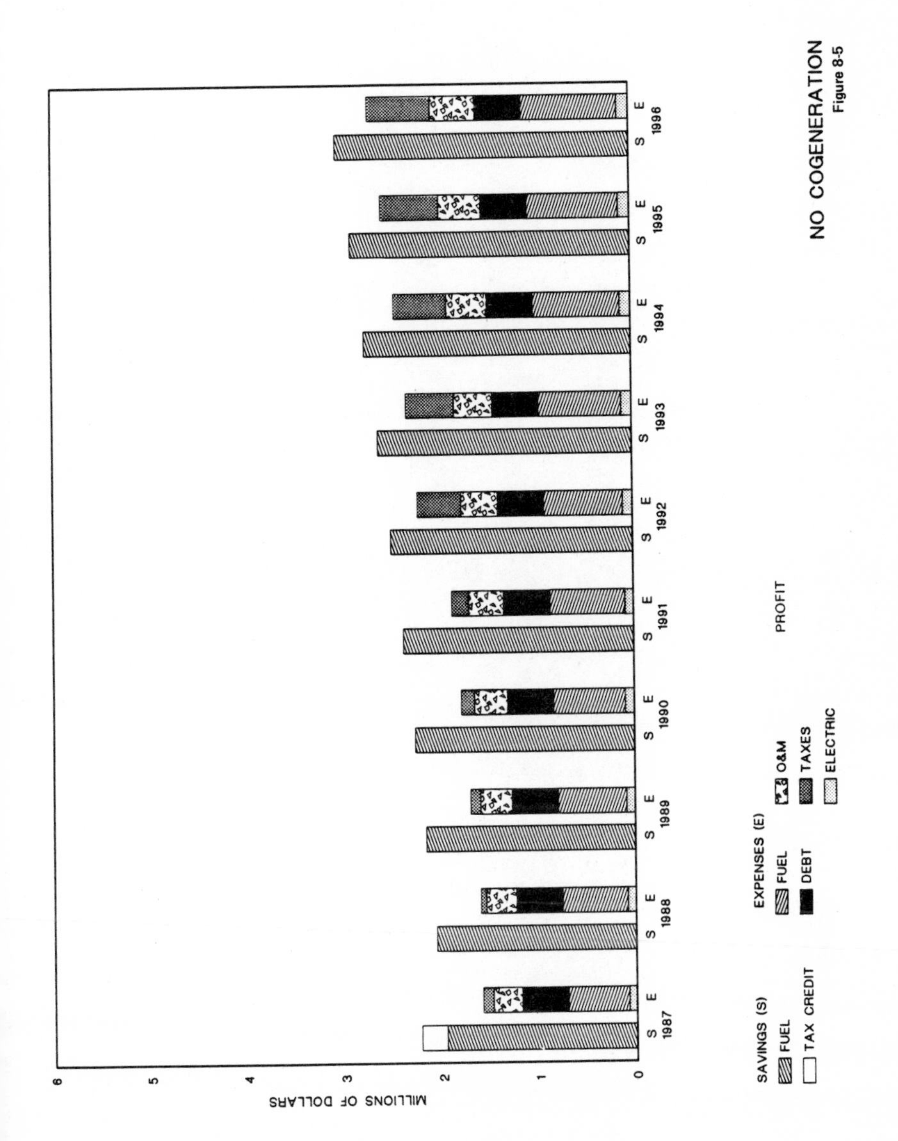

NO COGENERATION
Figure 8-5

CHAPTER 9

Exxon Chemical's Coal- Fired Combined-Cycle Power Technology

John J. Guide, P.E.

INTRODUCTION

The higher efficiency advantage of combined-cycle power systems is well known for Industrial and Utility Plants. These systems combine gas turbines, waste heat boilers and steam turbines to maximize the efficiency of steam and power generation. The major disadvantage of these systems is that they require a premium fuel, normally natural gas, to be fired in the gas turbine. Efforts are currently underway, by others, to develop coal-fired, combined-cycle systems. Some of these are based on the coal conversion processes of gasification or liquefaction to produce suitable gas turbine fuels. Alternatives employ pressurized or atmospheric fluidized beds for the combustion of coal. All of these systems use new technology and are not presently economically competitive. The best coal-fired power technology continues to be the relatively inefficient straight condensing power cycle for the Utilities and the efficient, but limited, backpressure cycle for Industrial Cogeneration.

Exxon Chemical's CAT-PAC concept is a direct pulverized coal-fired combined-cycle which utilizes elements of proven technology. It involves the marriage of a conventional direct pulverized coal boiler radiant section with a convection section adapted from process furnace experience. In particular, it is an open-cycle, hot air turbine arrangement, with indirect heating of the air in the convection section of a pulverized coal-fired boiler. The turbine exhaust is then used as pre-heated combustion air for the boiler. The air coil heats the 150 psig air, from the standard gas turbine axial compressor to approximately 1750°F. The mechanical design and the strength of the convection section metallurgy has been proven in our process furnaces at similar pressures and temperatures. Corrosion rates from coal flue gases for similar materials in both the low and high temperature regions have been established by recently completed EPRI- and DOE-sponsored probe test programs. These programs show the suitability of the CAT-PAC air heater metallurgy for coal service. U.S. Patent No. 4,479,355 has been issued to Exxon covering the CAT-PAC system.

General Description of Utility Plant Application

In the CAT-PAC simplified Utility Plant arrangement, Figure 9-1, air is compressed in a gas turbine air compressor, then heated in the convection section of a coal-fired boiler to a temperature of 1700°F. The hot compressed air is expanded in the gas turbine to generate power, and exhausts into the coal boiler windbox as preheated combustion air. Radiant heat in the boiler, along with convective heat not required for the gas turbine air heating, produces high pressure, superheated steam. This steam is expanded in a conventional condensing turbine for power generation and also provides extraction steam for feedwater heating/deaerating. Therefore, CAT-PAC is an indirectly fired, open-cycle turbine integrated with a coal-steam power plant.

The improved efficiency of the CAT-PAC cycle results from the recovery of all gas turbine exhaust as preheated combustion air and the partial substitution of the more efficient gas turbine cycle for the condensing steam cycle. The steam components and coal combustion equipment can be operated at conventional high efficiency design points with regard to temperatures, pressures, and excess air rates.

Conceptual Design of a CAT-PAC Utility Plant

Figure 9-2 is a more detailed flow plan of the conceptual CAT-PAC design for a fully fired, combined-cycle utility plant. This plant generates 263 MW of net power at a heat rate of 8930 Btu (HHV)/kWh (38.2% HHV). This design represents a single fully fired configuration for a commercial plant today. Supplementary-fired (high excess air) lower heat rate designs will be discussed later in this chapter.

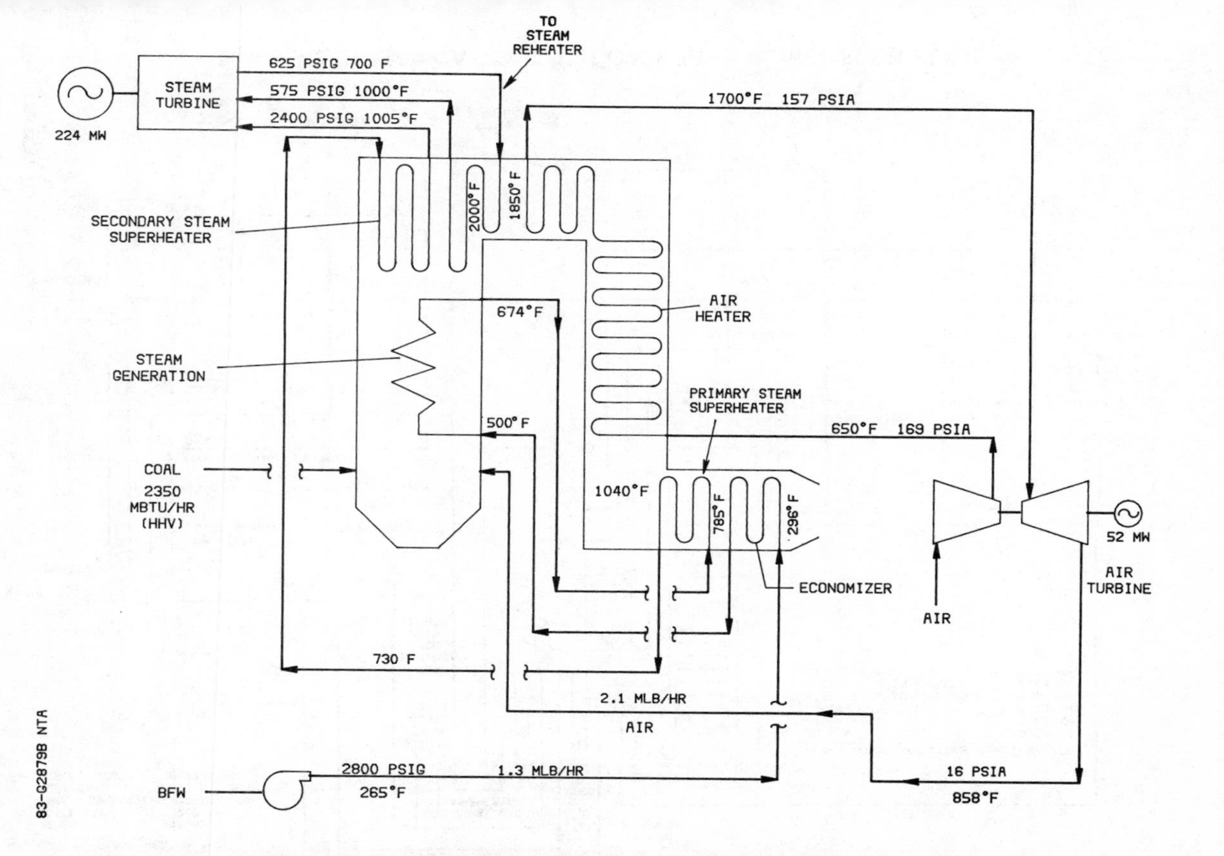

Figure 9-1. 263 MW CAT-PAC UTILITY POWER PLANT SIMPLIFIED FLOW DIAGRAM

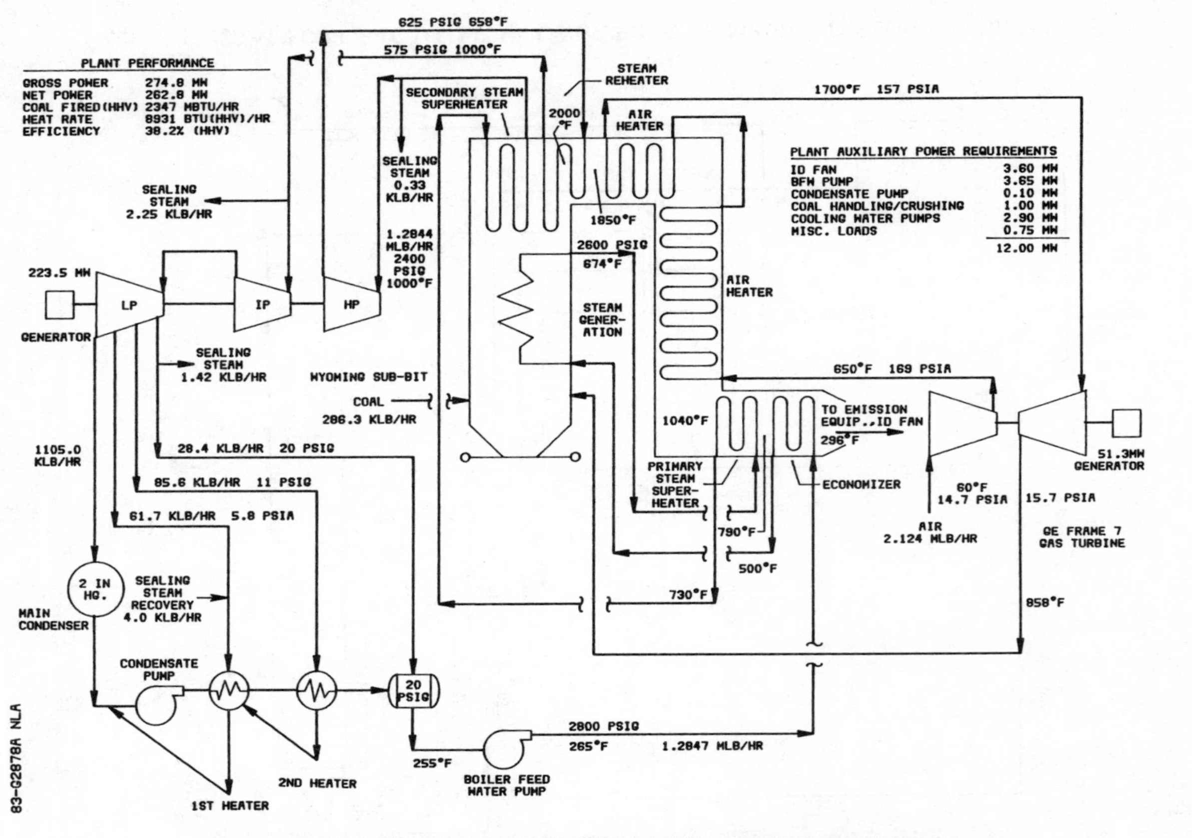

Figure 9-2. 263 MW CAT-?AC UTILITY POWER PLANT FLOW PLAN

Hot Air Turbine

In this design, 2.1 M lb/hr of air is compressed by an axial air compressor to 169 psia, 650°F. The compressed air is heated in the convection section of the coal boiler to 1700°F, and then expanded in the gas turbine to produce power. The exhaust air is ducted to the boiler windbox (and pulverizers) where it provides 100% of the boiler's required combustion air.

The air compressor has an 11.5 compression ratio with an 87% isentropic efficiency of compression and the expander has an 88% isentropic efficiency of expansion. These data correspond to typical performance of several off-the-shelf gas turbine generators (75 MW). Those designed with external combustors particularly lend themselves to this application.

The gas turbine combustors are not used, as all the required heat is provided by the boiler. Air leaves and returns to the gas turbine via ducting which was designed for about 50 ft/sec air velocity. Based on ducting and convection section design, pressure drop between the compressor and the expander is about 12 psi.

Boiler

Figure 9-3 is a simplified sketch of the boiler arrangement. Shown is a conventional pulverized coal-fired furnace system with natural circulation and with tangentially fired burners. Platens are used to perform portions of the steam superheat and the single steam reheat duties. The convection section provides air heating, economizer, superheating, and reheating duty. The heat absorbed in each section of the boiler is also shown.

The CAT-PAC boiler design concept shifts various steam service duties in order to free up high level convection duty for air service. This shifting is accomplished mainly by three steps:

1. Elimination of the combustion air heater frees up flue gas between about 800°F and 300°F for other uses.
2. Reducing economizer coil outlet temperature reduces economizer use of convective flue gases, freeing more convective flue gas for other uses.
3. Increased radiant platen use in superheater service frees up more convective flue gases.

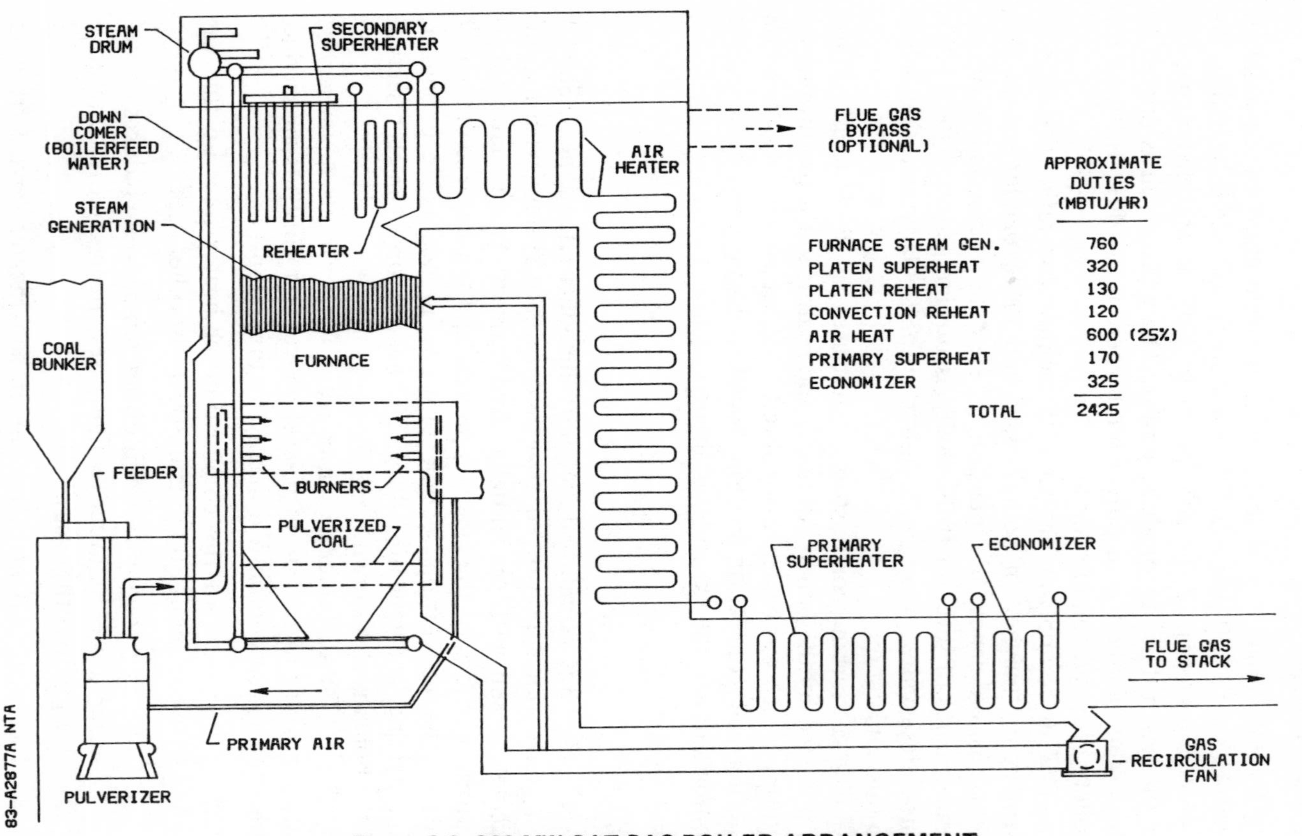

Figure 9-3. 263 MW CAT-PAC BOILER ARRANGEMENT

Roughly, the same radiant-to-convective heat duty ratio has been maintained for CAT-PAC versus a conventional utility boiler.

Air Heater

Figure 9-4 is a more detailed convection section layout. The air heater section has adequate spacing at a 50 ft/sec velocity. The overall heat transfer coefficient averages out to about 10 Btu/hr·ft^2·°F for the section. Hence, 238 kft^2 of bare 6-inch tubes on square pitch are provided. Layout of the vertically hung tubes on the top section consists of 16 passes with 4 by 32, or 128 tubes per pass. The side section, which uses horizontal tubes, consists of 11 passes with 8 by 20, or 160 tubes per pass. This layout ensures both adequate heat transfer and acceptable pressure drop. The total length of the air heater is about 200′ with sootblowers spaced throughout.

Four different metals are used for the air heater tubes. About 20% of the tubes are made of 2¼ CR which are used in the lowest temperature region with a maximum tube metal temperature of about 1080°F. The next 40% of the tubes use 304 SS up to a maximum tube metal temperature of about 1440°F. If some of these tubes, depending on coal selection, will be exposed to very corrosive molten alkali-iron trisulfates, they will require slight additives of aluminum, silicon and manganese to the commercial 304 composition and/or a magnesium zirconate coating. These options are detailed in the EPRI/FW Probe Test Program Final Report EPRI CS-3134 (July, 1983). An alternative design places some steam superheat duty in this region to avoid tube metal temperatures in this corrosive temperature window. Incoloy 800 H tubing is used for the next 20% up to tube metal temperatures of about 1620°F. The remaining tubes in the highest temperature region with tube metal temperature up to 1800°F, use HP-modified material. This basic 25 CR-35Ni material was bracketed by lower grade and better metallurgy in the Probe Test Program carried out by DOE/C-E and published in the Final Report DOE-AC01-76ET1050 (Dec., 1980). This choice of metallurgy gives due consideration to stress and corrosion factors with reasonable wall thicknesses at the high tube metal temperatures of the air heater. These materials, with these tube diameters and wall thicknesses, have been standard in Exxon fired heaters at approximately equivalent stress values. Placement of the air coil in an area below ash fusion temperature avoids slagging and also helps inhibit

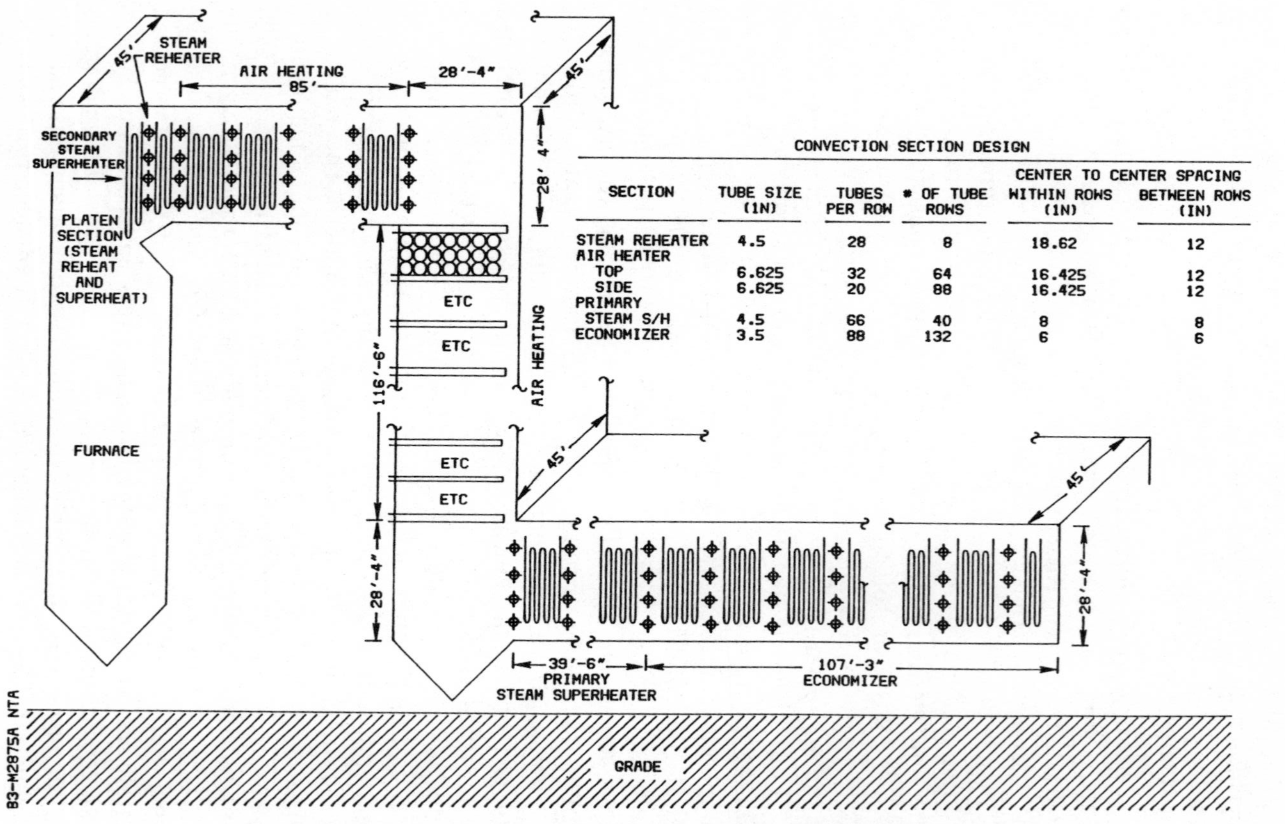

CONVECTION SECTION DESIGN

SECTION	TUBE SIZE (IN)	TUBES PER ROW	# OF TUBE ROWS	CENTER TO CENTER SPACING WITHIN ROWS (IN)	CENTER TO CENTER SPACING BETWEEN ROWS (IN)
STEAM REHEATER	4.5	28	8	18.62	12
AIR HEATER					
TOP	6.625	32	64	16.425	12
SIDE	6.625	20	88	16.425	12
PRIMARY STEAM S/H	4.5	66	40	8	8
ECONOMIZER	3.5	88	132	6	6

Figure 9-4. 263 MW CAT-PAC BOILER CONVECTION SECTION LAYOUT

tube and tube support corrosion. The air heater tube support system will be a mixture of fired heater and boiler technology, but primarily fired heater technology.

CAT-PAC High Efficiency Design

The CAT-PAC utility plant conceptual design described previously represents a single fully fired configuration for a commercial plant today. More efficient designs for the same 2400 psig, single reheat steam cycle boiler would increase gas turbine penetration. Well-known methods that would increase the gas turbine penetration and lower the heat rate include flue gas recirculation, high excess air (about 120%) operation, and higher air turbine initial conditions.

Higher pressure and temperature (2200°F for modern gas turbines) operation of the expander inlet can be obtained by improving tube metallurgy or by trim firing (possibly coal gas) in the gas turbine combustors. These steps would increase thermodynamic cycle efficiency. Using these techniques, it is expected that a heat rate of 7750 Btu (HHV)/kWh (44% HHV) could be developed for CAT-PAC. Figure 9-5 shows the positive effect on CAT-PAC cycle efficiency for higher excess air and higher turbine inlet temperature operation. These efficiency improvements also apply to the Industrial Cogeneration application which will be covered next.

CAT-PAC Industrial Application (Cogeneration)

The CAT-PAC scheme is not limited to utility boilers. Industrial steam and power systems can also benefit from this technology. A simple example of this cogeneration application is shown in Figure 9-6. We also did a more in-depth, side-by-side comparison of a base case and a CAT-PAC case for an 1860 psig industrial boiler. The base case generated about 77 MW of power and exported about 1.6 Mlb/hr of 600 psig steam. This cogeneration plant operated at 85% (HHV) cycle efficiency. Increasing power output from coal without CAT-PAC for this plant can be done only at the expense of efficiency. At best, a 30% (HHV) efficient condensing cycle would be used to produce the additional power. CAT-PAC can increase the plant's power generation by 77% and keep the same high 85% (HHV) cycle efficiency. This capability is very attractive for industrial application.

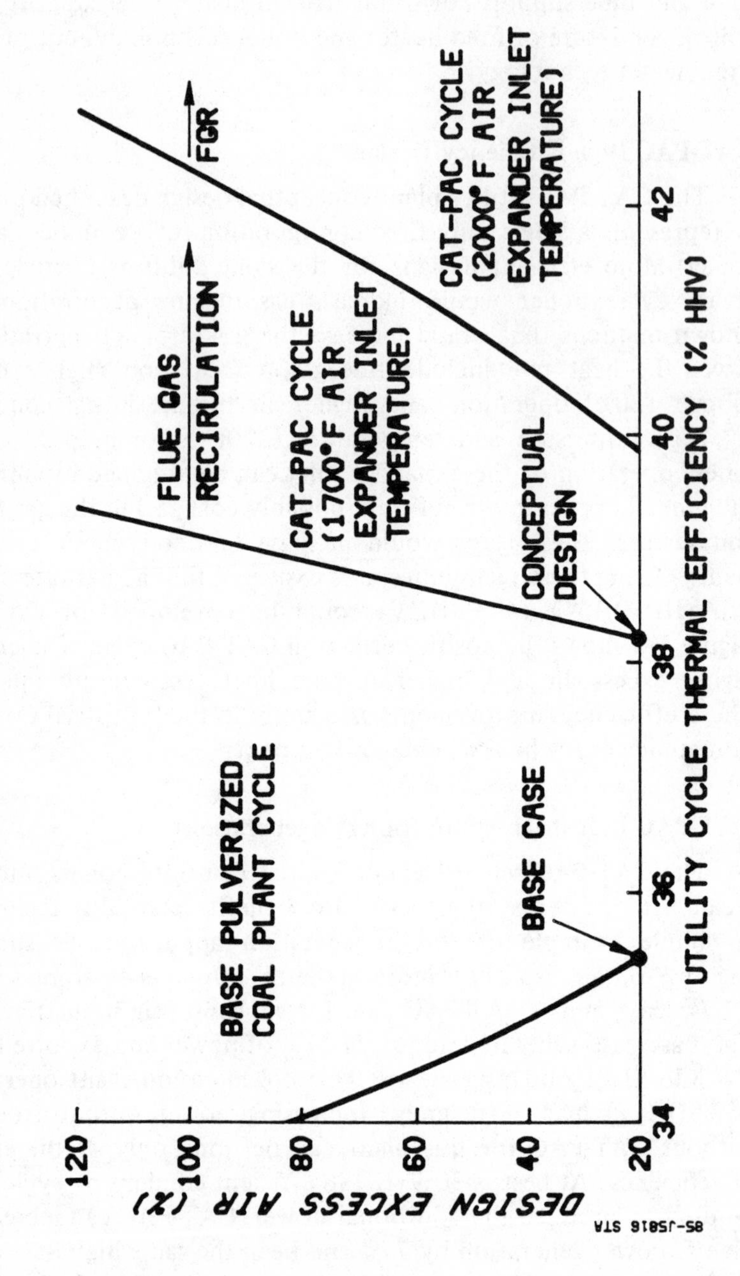

Figure 9-5. UTILITY CYCLE THERMAL EFFICIENCY (%HHV)

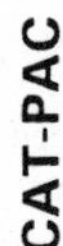

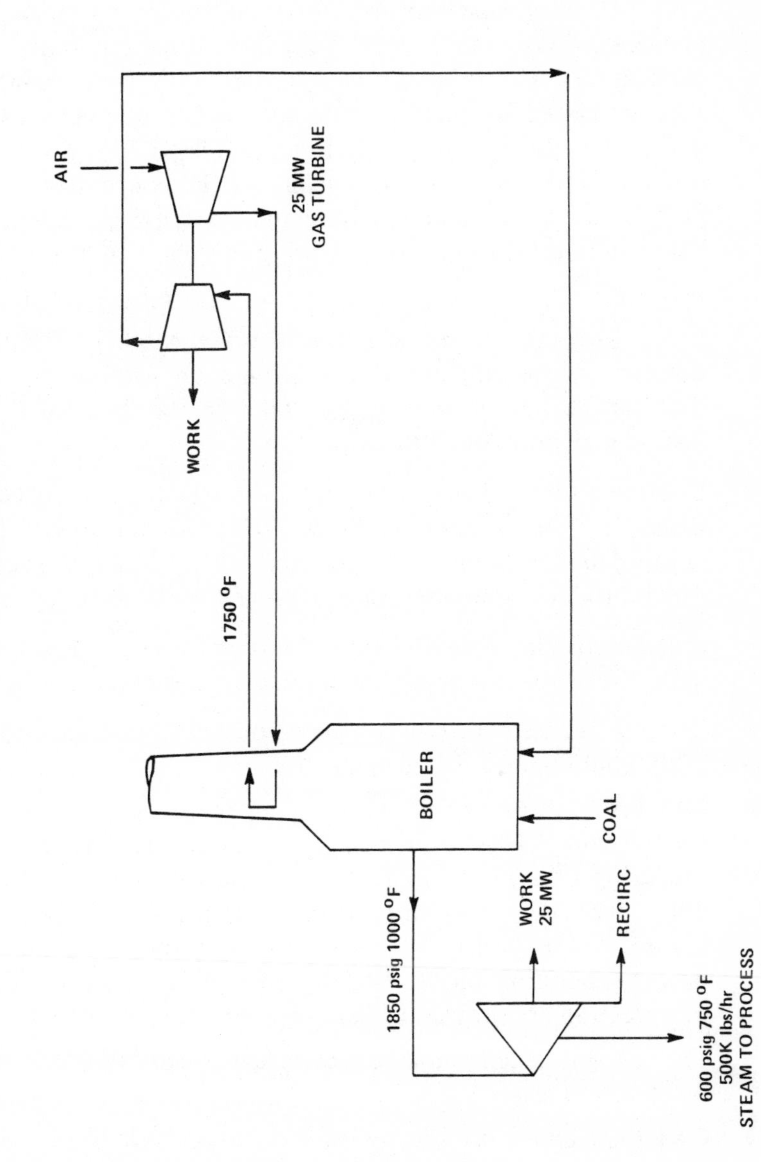

Figure 9-6. CAT-PAC COGENERATION APPLICATION

Application Design Considerations of CAT-PAC

- CAT-PAC is a *modular generation technology* available in practically any size power plant for utility or industrial applications. Utility plant net power requirements can be accommodated by increasing/decreasing the number and/or size of the gas turbine used. For example, a 525 MW plant could be built by doubling the size of the conceptual CAT-PAC boiler design and by using two of these gas turbines. Using four gas turbines would bring the plant to 1050 MW.
- *Fuel flexibility* is a strength of the CAT-PAC combined cycle. The mechanism which achieves flexibility is simply the indirect heating of the gas turbine air. Almost any fuel that can be burned in a utility boiler can be used for CAT-PAC, e.g., lignite, low Btu gas, etc.
- *Existing utility plants could repower* (or convert to coal) with CAT-PAC to produce up to 50% more net power. The existing boiler could be replaced with a CAT-PAC boiler which produces the same steam flow and conditions.
- *Existing industrial plants could repower* (or convert to coal) with CAT-PAC to nearly double their cogenerative power.
- *Existing gas turbine power plants,* both Utilities and Industrials, could be retrofitted with CAT-PAC.
- CAT-PAC is compatible with and enhances *compressed air energy storage* (CAES) systems. CAT-PAC would not require the firing of a premium fuel (one-third premium fuel firing duty over the two-thirds stored compression duty) (see Figure 9-7) as is usually done in today's designs during the power generation mode. Peaking and energy storing modes of operation are shown in Figure 9-7.
- The design is adaptable to different *steam pressure and temperature levels.*
- CAT-PAC *emissions are reduced* on a per MW basis versus a base case pulverized coal plant. This is because CAT-PAC produces more power for less fuel. Emissions are controlled using conventional FGD scrubbers and bag filters or electrostatic precipitators. CAT-PAC's reduced cooling water

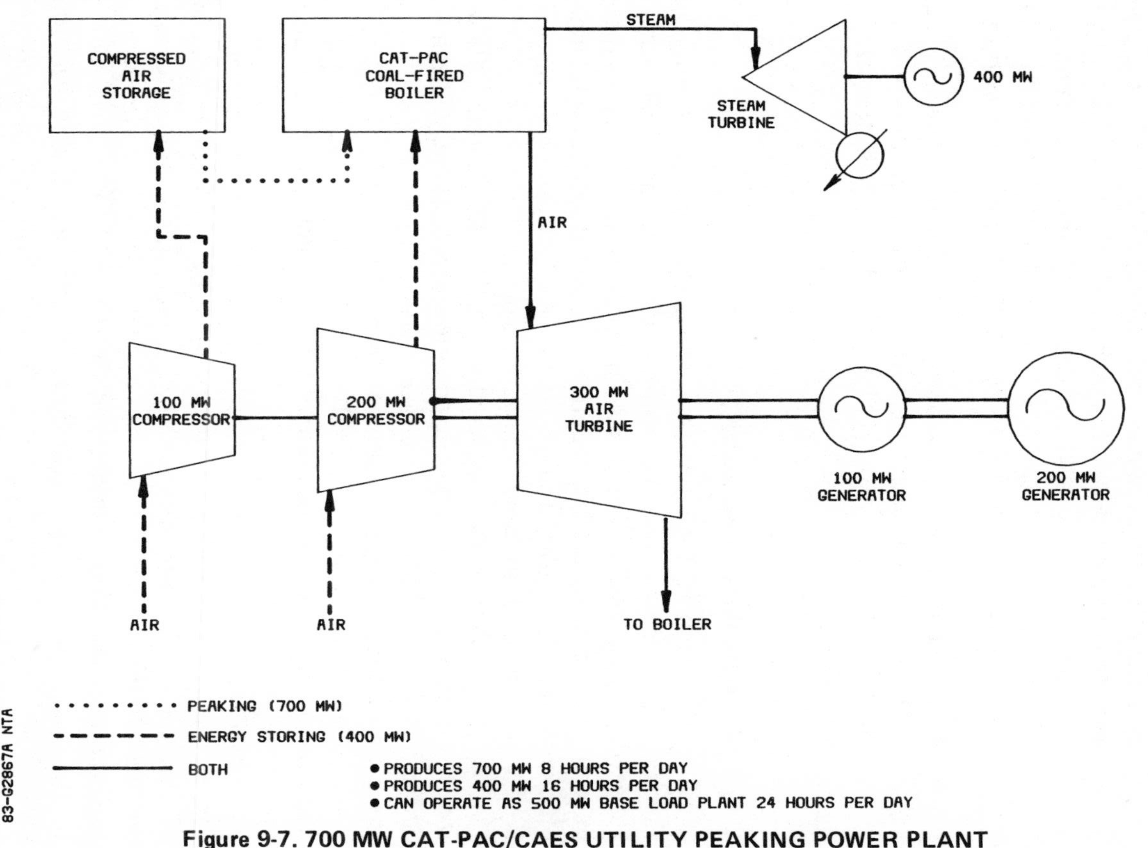

Figure 9-7. 700 MW CAT-PAC/CAES UTILITY PEAKING POWER PLANT

requirements and reduced steam rate will significantly reduce the plant's water requirements and losses.

- It is expected that the plant will have *as good a service factor* as a base loaded pulverized coal plant. Unlike a direct fired gas turbine, the indirectly heated gas turbine operates on air. This means that significant downtime due to the combustion process and the degradation of turbine blading will be avoided.

- The overall plant is *no more complex* than a conventional coal plant plus peaking gas turbines. The plant does not involve equipment or operations new to the Utilities or Industrials.

CAT-PAC Technology in Perspective

The discussion of CAT-PAC has so far compared CAT-PAC to the state-of-the-art pulverized coal-fired utility and industrial plants. However, much research and development work is on-going to develop new state-of-the-art coal-fired combined cycles with high efficiency and low cost/strategic fuels. This section will discuss CAT-PAC in relation to these other technologies.

The two technologies considered are coal gasification and fluidized bed combustion (atmospheric and pressurized) combined cycles. These technologies have an advantage in easier sulfur removal versus the conventional flue gas desulfurization process employed by CAT-PAC. However, based on an overall comparison, CAT-PAC has many major advantages over these competitive arrangements and can also meet required environmental specifications.

CAT-PAC links elements of proven technology as shown in Figure 9-8 and therefore, is here today. The competitive designs of coal gasification and fluidized bed combustion combined cycles are less efficient, need major technical developments, have shorter run lengths, are more complicated or require higher investment. In summary, all of these "coal-fired gas turbines" should be progressed as the economically optimum technology will depend on specific applications.

The 263 MW CAT-PAC Utility Power Plant investment was estimated to be the same as the Base Case Coal Plant.

RADIANT SECTION	– Existing pulverized coal-fired boiler technology using membrane waterwalls etc.
CONVECTION SECTION:	
STRENGTH	– Existing chemical reformer and steam cracking furnace technology with similar pressures and temperatures and metallurgy
COAL CORROSION	– Extensive probe test programs recently completed by EPRI/FW and DOE/C-E established corrosion rates which show the suitability of CAT-PAC air heater metallurgy for this application
ENVIRONMENTAL	– Existing pulverized coal-fired boiler technology including latest advancements in scrubbers, precipitators, limestone injection, etc.

Figure 9-8. CAT-PAC USES PROVEN TECHNOLOGY

CHAPTER 10

Cogeneration at Small Plants: Design Using Reciprocating Engines And Absorption Chiller

Joseph R. Wagner

INTRODUCTION

The last dozen years have seen a surge of interest in industrial cogeneration systems (i.e., systems which simultaneously produce heat and electricity). This surge began with the 1973-74 Energy Crisis, and was in response to dramatic increases in the cost of fossil fuels. Despite the stabilization of fossil fuel prices in the 1980s, interest in cogeneration has continued to grow. This can be credited to a number of factors, including the demonstrated success of cogeneration projects installed during the last decade and the increasing availability of proven equipment to meet the needs of different types of industrial facilities.

Despite the increasing variety of options, there are still applications for which suitable equipment is difficult to find. Small-scale cogeneration is one such application. For the purpose of this chapter, small scale means that the user wants a continuous output of, at most, 1 MW of cogenerated electricity. The plant's total electrical demand might in fact be much larger.

There are three basic strategies for cogeneration, categorized by the primary fuel-using device. These are:

- Gas turbine
- Steam boiler
- Reciprocating engine.

Plant operators often find gas turbines unsuitable for small-scale applications. One important factor in this determination is the generally low fuel-to-shaft efficiency of gas turbines in this size range.

The percentage of fuel energy that is converted to electricity is significantly lower than that of larger gas turbines. This can undermine project economics. A steam boiler system, with a backpressure turbine for electrical generation, likewise generally converts a relatively small portion of the fuel energy into electricity. This is especially true if the boiler operates at the modest pressure levels typical of small plants.

For these reasons the first two strategies above do not meet the requirements of all plants. This chapter proposes an alternative design based on the third strategy, namely, reciprocating engines.

PROS AND CONS OF RECIPROCATING ENGINES

Relative to the prime movers used in other cogeneration strategies, reciprocating engines offer the following advantages:

- Low capital cost, on a dollar per shaft horsepower basis
- High fuel-to-shaft efficiency
- Good part-load efficiency
- Good load following
- No need for gas compressor (as is the case with gas turbines) when fueled with natural gas
- Speed range allows direct drive of generators
- Many suppliers and wide selection in the $\leq$1MW size range

Disadvantages include relatively high maintenance costs, on a ¢/kWh basis. This tends to be the case if the engine is fueled with natural gas. Such engines were designed initially to burn diesel fuel (i.e., distillate oil), and have been converted to natural gas by the addition of spark plugs and other modifications. Depending on the quality of the modifications, problems can arise in a number of areas, including the spark plugs and exhaust valves. Aside from the expense of maintaining these components, unscheduled outages can occur and, in some cases, can severely affect project economics.

Another disadvantage of reciprocating engines is that they are poor steam generators when compared to gas turbines. Virtually all waste heat leaves a gas turbine in the form of high-temperature gas, and can be readily converted to process steam at pressures typical of the main steam header at a small plant (e.g., 125 psig). Reciprocating engines, on the other hand, reject about half their waste heat to the cooling water circulating through their engine blocks. This so-called jacket water exits the block at only about 200°F. Consequently, the most successful cogeneration applications for these engines involve situations where the user has a direct need for hot water.

SYSTEM DESIGN APPROACH

Choice of fuel is the first issue confronted by an aspiring cogenerator. The motivation to cogenerate is a function of the difference between the price of purchased electricity and the price of the fuel used by the cogeneration system. The greater this difference, the greater the savings potential. This dictates using the cheapest fuel (on a $/MMBtu basis) that the system can accommodate without increasing capital or maintenance costs.

For reciprocating engines in the United States, natural gas is the fuel of preference. As compared to the diesel engine from which it is derived, a natural gas engine will be derated, somewhat less efficient, more expensive to maintain, and more expensive on a $/kW basis. Yet these disadvantages are frequently more than offset by the low price of natural gas relative to diesel fuel. This is especially true where the gas utility is providing preferential rates, or where users produce their own gas. Since small-scale cogeneration is a developing market, these high-potential applications should be addressed first. Thus, the present design effort addresses engines fueled with natural gas.

Aside from fuel type, choices must be made regarding other design options, as indicated in Figure 10-1. The shaft power can be used to drive either a variable-speed electric generator, which would require a dedicated circuit to a variable frequency load, or a fixed-speed generator, or to drive mechanical loads such as compressors, pumps, and fans. The present system was designed to drive a fixed-speed electric generator, since 60-Hz electricity can be used directly throughout a plant, with a minimum of site engineering. This makes the system easier to retrofit to a large population of plants.

With this approach, a number of grid interconnect schemes are possible. Direct connection to the grid could be maintained at all times, allowing the plant to buy or sell power. Alternatively, the grid

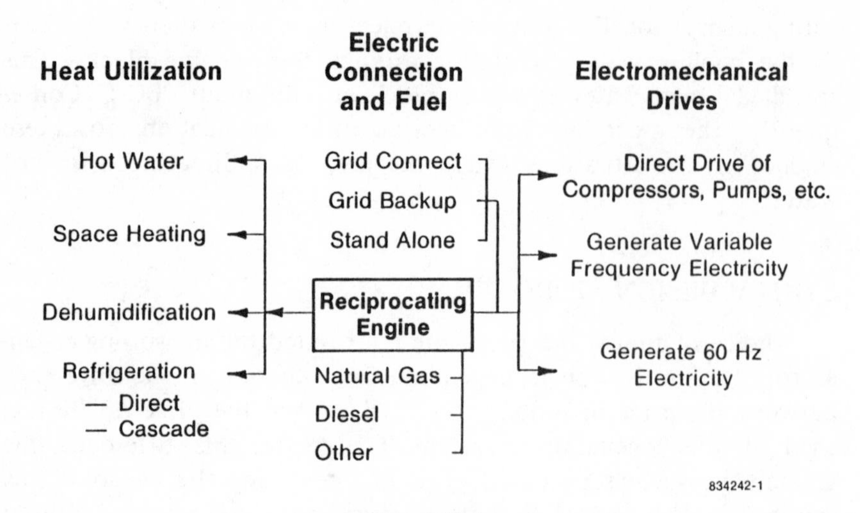

Figure 10-1
SYSTEM DESIGN OPTIONS

could be used only for backup. That is, the cogeneration system would supply all plant power. If the cogeneration system failed, or could not satisfy the load, the system would shut down and the plant would temporarily reconnect to the grid to buy power. Another approach is the stand-alone option. This requires synchronous generators, a degree of overcapacity or loadshedding ability, and a high level of redundancy throughout the cogeneration system.

The local utility's rate schedule, interconnection specifications, and general attitude towards cogeneration dictate which approach is appropriate. The present system is designed for direct grid connection, and economic calculations are based on the assumption that the plant's base electricity consumption exceeds cogeneration capacity, such that the plant would always be buying power from the grid, and the cogeneration system would always be loaded heavily. This situation is economically equivalent to a case in which cogenerated electricity exceeds plant demand, and the utility buys the excess at a price equal to the utility's selling price.

Another decision necessary regarding basic design concerns the engine's waste heat. As noted above, a reciprocating engine is economically most attractive in situations where the hot jacket water can be used directly. Unfortunately, most plants have limited use for any new source of hot water. Applications such as domestic hot water, space heating, and boiler water preheating are usually small, seasonal, or are easily satisfied by recovering waste heat from existing plant equipment. Potentially, the jacket water could be sent to an absorption chiller, but the coefficient of performance (COP) for such units is low, and the resulting refrigeration tonnage is small.

This often leaves the engine's exhaust gas as the only usable form of heat. It can be used to generate a small amount of steam, or possibly can be linked to a desiccant dehumidifier to help reduce the load on the plant's refrigeration and air conditioning system. Another option is to send the exhaust gas to an absorption chiller, where it can be used to produce chilled water. The COP is typically significantly greater than for a chiller driven by hot water. The chilled water can be used directly for refrigeration, air conditioning, or process cooling, or could be used as part of a refrigeration cascade, as will be described.

The chiller option was selected for the present design because it was judged applicable to the greatest number of plants. A contributing factor to this decision was the identification of a chiller that could simultaneously use both the exhaust gas and the jacket water. Thus, this strategy allows use of all the engine's waste heat and converts it into a form deemed most likely to be usable at a typical plant.

To further increase the system's applicability, it is designed as a packaged module. This reduces site engineering and installation costs. It also allows better integration of system components. The final system is designed to be a factory-assembled, standardized product capable of achieving economies accruing from volume production.

System Layout

Figure 10-2 is a schematic showing the functional relationship between basic system components. The system shown contains three engines, each of which drives an electric generator. Multiple engines were specified to avoid creating demand peaks. A single, larger engine-generator can do the work of three smaller units, but when a single, large engine fails or is shut off for routine servicing, the utility sees this as a large demand spike. Depending on the local rate schedule, this can have a significant effect on project economics,

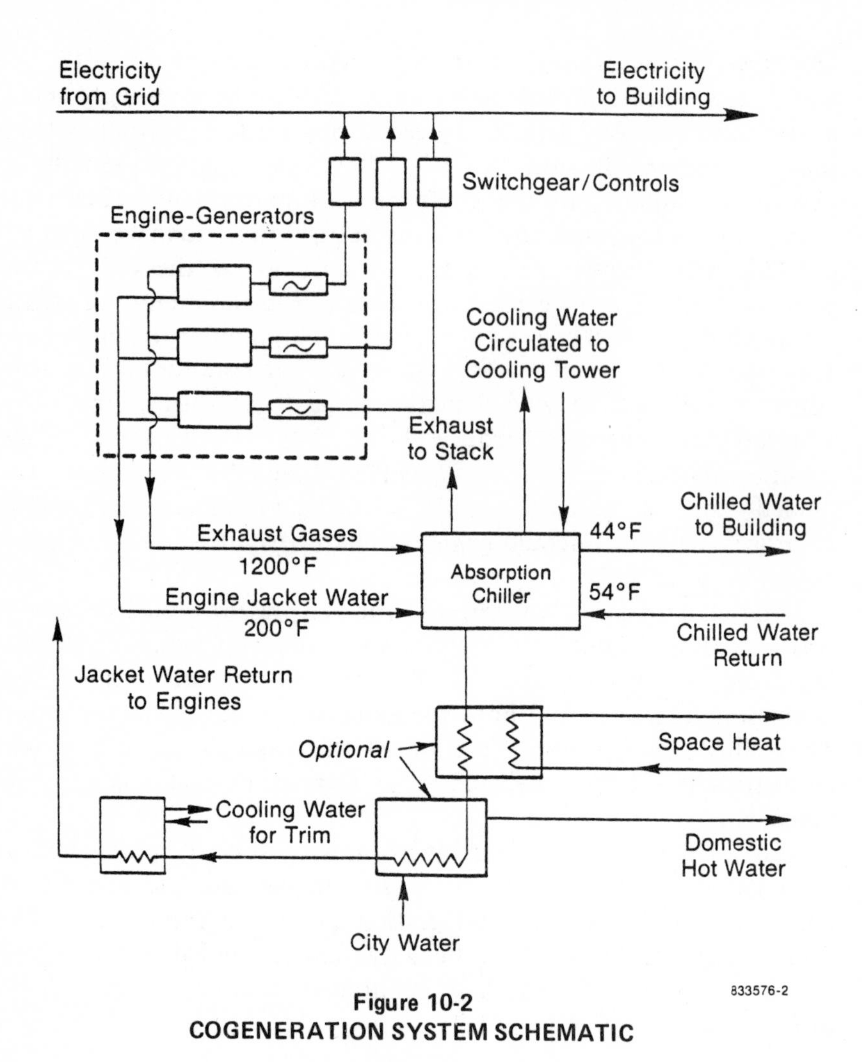

Figure 10-2
COGENERATION SYSTEM SCHEMATIC

since demand charges in some service territories are very high. Another advantage of the multi-engine approach is that it is more adaptable to stand-alone strategies, where redundancy and ability to modulate system output are more important.

The actual physical dimensions of the proposed system are indicated in Figure 10-3. This layout depicts a three-engine skid that can be transported on a standard 45-ft trailer. Gross vehicle weight limitations would probably require that the engine-generators

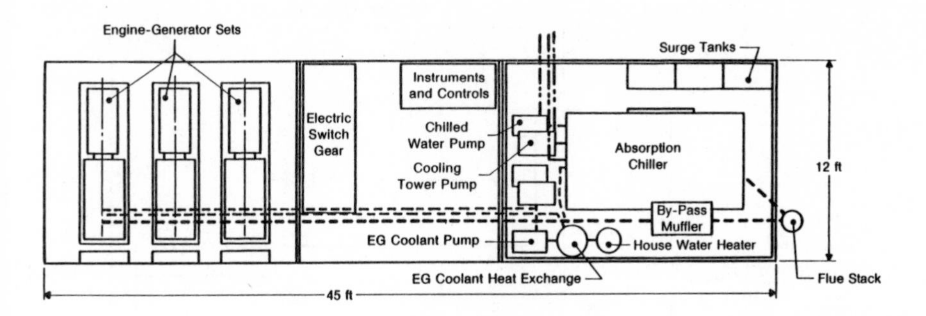

Figure 10-3
PLAN VIEW: EQUIPMENT LAYOUT

be transported separately. The engine generators are designed for easy changeout, primarily to satisfy servicing requirements. This feature also allows simplified installation.

The total basic continuous-duty electric generating capacity for all three engines is approximately 418kW. Depending on the components selected, the basic layout's electric generation capacity can be designed to be as low as 279 kW or as high as 626 kW. The higher capacity figure would require turbocharged engines. Capacities of 1 MW or more could be achieved by selecting engines having larger displacements, but this would require at least two skids, and would tend to violate the concept of a packaged system.

Refrigeration Cascade

As shown in Figure 10-2, the system is designed to produce chilled water at about 44°F. This is the approximate lower limit of the refrigeration temperature achievable with a lithium-bromide absorption chiller. Presently, this is the only chiller technology identified as commercially available and capable of satisfying various technical constraints. This would seem to limit application of this system to plants having refrigeration loads of 44°F and higher. In fact, this is not so, since a refrigeration cascade can be employed.

Figures 10-4 and 10-5 provide a simple example of a refrigeration cascade applied as a retrofit. Figure 10-4 shows a plant refrigeration system prior to retrofit. Refrigerant R-12 (a type of Freon™) is arbitrarily assumed as the working fluid in a system maintaining

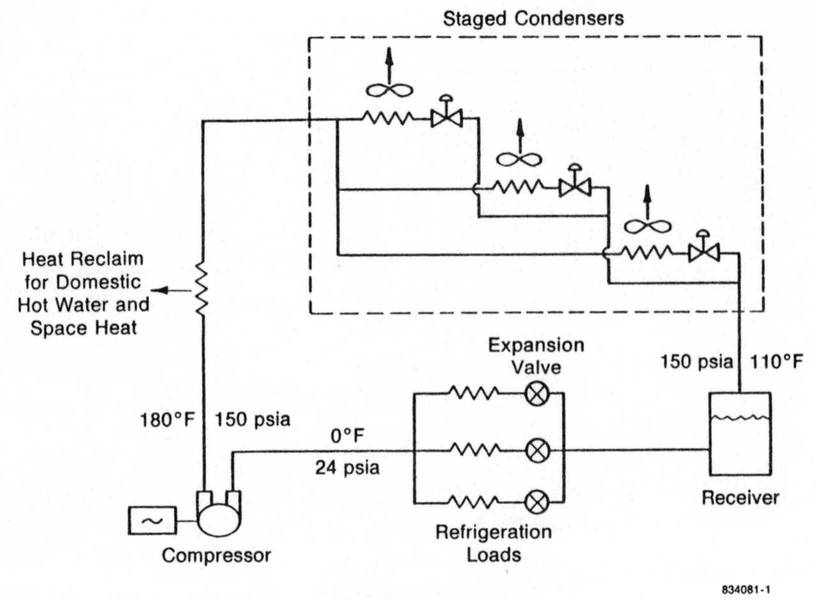

Figure 10-4
EXISTING REFRIGERATION SYSTEM
(R-12 Values Used for Illustration)

various refrigeration loads at 0°F (i.e., 0°F suction temperature). A compressor takes in the 24 psia Freon vapor exiting the heat exchanger(s) in the refrigerated space(s), and compresses this vapor to 150 psia.

This exhaust pressure is a function of the operation of the condensers employed in the refrigeration system. The performance of the condensers is, in turn, governed by ambient weather conditions. On a hot day the condensers must work harder to extract the necessary amount of heat from the Freon vapor, so it will condense and allow the refrigeration cycle to function. If condensation does not occur fast enough, pressure increases on the condenser side of the system.

The overall system is set up so that the expansion valves have a relatively uniform upstream pressure. This requires that the system be set up for "worst day" conditions, and results in the compressors having to contend with a relatively high exhaust pressure, regardless of weather conditions. This exhaust pressure is the pressure that the

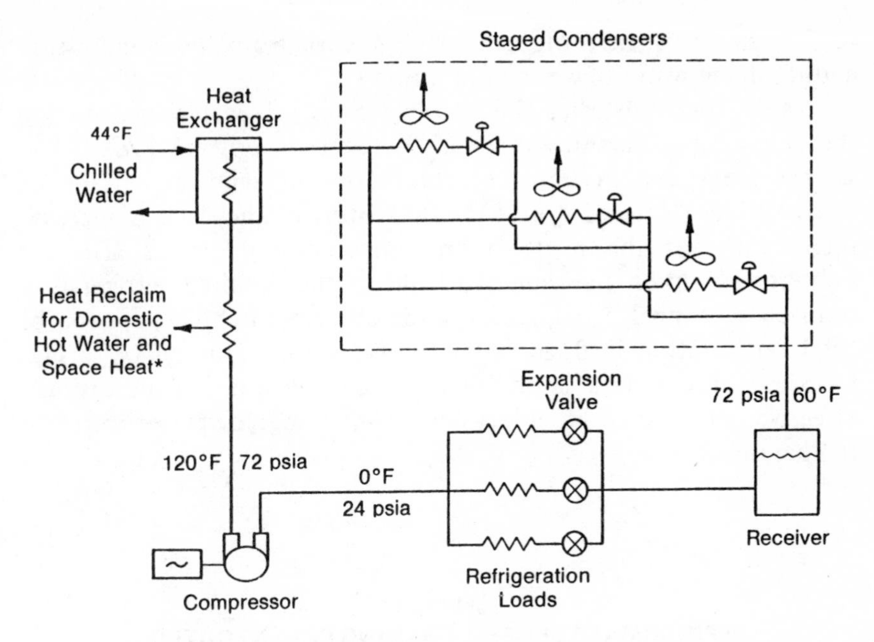

Figure 10-5
REFRIGERATION SYSTEM AFTER RETROFIT
(R-12 Values Used for Illustration)

condensers can ensure will not be exceeded, unless weather conditions exceed the "worst day" design conditions. As shown in the example, the worst day is anticipated to be one on which the condensing temperature would be 110°F, with the compressors working at full capacity. This is a typical design temperature in the U.S. Condenser sections are turned on or off to maintain this temperature, and hence to maintain the fixed pressure of 150 psia.

Figure 10-5 shows how this situation would change by instituting a refrigeration cascade. A new heat exchanger would be installed to allow condensation to be performed by 44°F chilled water, as supplied by a lithium-bromide chiller. The chilled water temperature is not subject to ambient conditions, and would allow readjustment of the refrigeration system. The example shows this readjustment resulting in a new condensing design temperature of 60°F, which reduces the controlled pressure on the condenser side of the system to 72 psia. This, in turn, requires less work from the compressors and results in lower power consumption.

Table 10-1 illustrates the power savings achievable by reducing the condensing temperature to 60°F versus an original 108°F. The savings are stated in terms of compressor horsepower per ton of refrigeration (Note: 1 ton = 12,000 Btu/h). The savings increase significantly as the refrigeration temperature decreases. The far right-hand column in Table 10-1 shows the required refrigeration tonnage that must be supplied by the absorption chiller, per ton of total refrigeration load, in order for the stated compressor savings to occur. These data imply that the chilled water is increasingly effective when used in a cascade maintaining lower refrigeration temperatures.

Table 10-1
APPROXIMATE BENEFIT OF USING COGENERATED CHILLED WATER TO REDUCE REFRIGERATION LOADS

Evaporator Temp. (°F)	Horsepower/Ton vs Condensing Temperature			Chiller Tonnage/Ton Refrigeration
	108°F	60°F	Savings	
40	1.27	0.33	0.94	1.07
30	1.54	0.51	1.03	1.11
20	1.84	0.71	1.13	1.15
10	2.17	0.93	1.24	1.20
0	2.55	1.18	1.37	1.25
-10	2.98	1.45	1.53	1.31
-20	3.48	1.74	1.74	1.37
-30	4.05	2.08	1.97	1.44
-40	4.73	2.45	2.28	1.52

*Suction temperature

834176

Component Selection

A variety of reciprocating engines is available for incorporation into the basic system design. Table 10-2 shows examples of commercially available engines, along with approximate technical characteristics. For illustration purposes, the Cummins G-855 is assumed to be the prime mover for the basic system described in this chapter. It has been selected primarily because it exhibits high efficiency relative to other engines in the approximate 200-hp size class. The 200-hp size class was selected as being consistent with the multi-engine strategy and with the desired overall electrical and thermal output of the system.

Table 10-3 provides additional detail on the mechanical and thermal output of the Cummins G-855. The right-hand column indicates the tons of refrigeration produced by a lithium-bromide chiller driven by the engine's waste heat. Note that these values are

Table 10-2
COMPARISON OF GAS ENGINE EFFICIENCIES
(Engines in Nominal 200-hp Class)

Parameter	Cummins G-855	Caterpillar 3306		Waukesha F1905GR
Displacement (in.^{3})	855	638	638	1905
Aspiration*	N	N	T	N
Max. Continuous-Duty Horsepower	200	145	220	240
Speed (rpm)	1800	1800	1800	1200
Fuel-to-Shaft Efficiency:**				
100% Load	0.311	0.297	0.291	0.259
75% Load	0.288	0.280	0.278	0.242
50% Load	0.250	0.246	0.247	0.207

834177

*N = Natural; T = Turbocharged

**On a higher heating value basis (1028 Btu/scf)

for only one engine. The design specifies that the output of multiple engines will go to a single chiller. Therefore, the values in Table 10-3 must be multiplied by the number of engines to show total system performance.

Table 10-3
THERMAL ENERGY CHARACTERISTICS OF CUMMINS G-855 GAS ENGINE (12:1 Compression Ratio, 1800 rpm)

	Energy Split	Equivalent Tons of Cooling
Mechanical Efficiency (%)	31.1	—
Jacket Water (10^3 Btu/h)		
100% Load	450.0	21.4
75% Load	390.4	18.5
50% Load	314.6	14.9
Exhaust* (10^3 Btu/h)		
100% Load	358.3	32.3
75% Load	293.8	26.5
50% Load	242.8	21.9
Total Tons of Cooling Generated by Chiller at 100% Load	—	53.7

852193

*Based upon reducing exhaust temperature to 375°F

The chilled water tonnage values are based on the performance characteristics of a Hitachi Paraflow™ chiller. To date this is the only commercially available chiller identified for this application in the U.S. This unit can use engine jacket water and exhaust gas simultaneously, and overall can provide an estimated 53.7 tons of refrigeration using the waste heat from one Cummins G-855. It has been suggested that the smallest Hitachi unit is, in fact, oversized for this application, but this does not affect performance estimates.

The chiller produces significantly more tonnage from exhaust gas than from jacket water (see Figure 10-6), even though there is more waste heat in jacket water than in exhaust gas. This is because the chiller COP increases with increasing engine waste heat temperature, and the exhaust gas is significantly hotter than the jacket water.

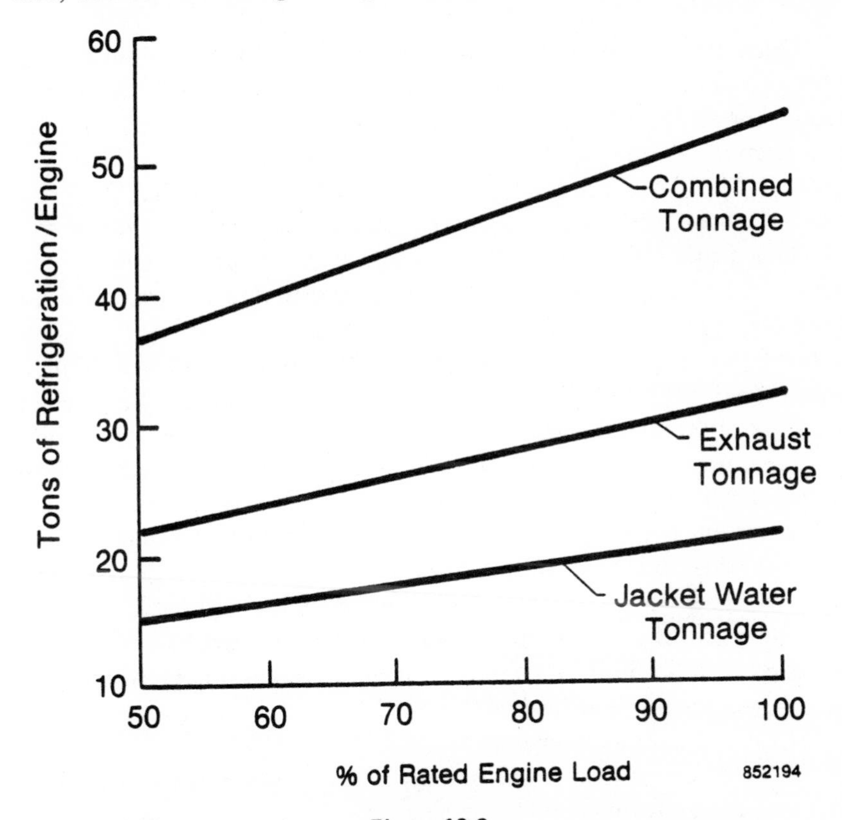

Figure 10-6
ABSORPTION CHILLER PERFORMANCE USING CUMMINS G-855 GAS ENGINE (12:1 Compression Ratio, 1800 rpm)

ECONOMICS

To determine the economic viability of the proposed system, an estimate of potential user benefits was developed and calculated under the following economic assumptions:

- System capacity: 418-kW electricity; 160-tons chilled water
- All electricity produced is used on site
- No stand-alone capability
- 80% load factor (average = 334 kW, 129 tons)
- Maintenance: 1¢/kWh produced ($29,000/yr)
- Average fuel-to-electricity efficiency: 28.8%
- Average chiller COP: 0.80 (combined exhaust and jacket water).

Once these assumptions are fixed, user benefits become a function of energy prices. Using a 2-year simple payback as a "hurdle rate," the minimum level of annual user benefits required to make the system economically attractive is equal to the installed cost divided by two. Conversely, the maximum allowable installed cost that a user can tolerate is equal to estimated annual savings times two.

This principle, which presumes the user requires a simple payback of 2 years or less, has been used to generate Figure 10-7. For example, Figure 10-7 shows that if electricity costs 8¢/kWh and natural gas is available at $4/MMBtu, the user can spend at most about $310,000 and still achieve a 2-year payback on the system. The $310,000 includes all equipment and installation costs.

Superimposed on this figure are lines showing anticipated installed costs. The "retrofit" line shows the estimated installed cost if the system is assumed to be equipped with its own enclosure and to bear the full cost of all interconnections with existing plant systems. The "new construction" line shows the estimated installed cost assuming that the system is incorporated as part of a new plant, or as part of a plant expansion or modernization, in which case many cost savings can be realized in the areas of enclosures, piping, controls, and heat exchangers.

The result is that even in the "new construction" scenario, system economics will likely be significantly attractive only in those areas of the country having relatively high electric rates and low-to-moderate natural gas rates. Nonetheless, these areas exist and are

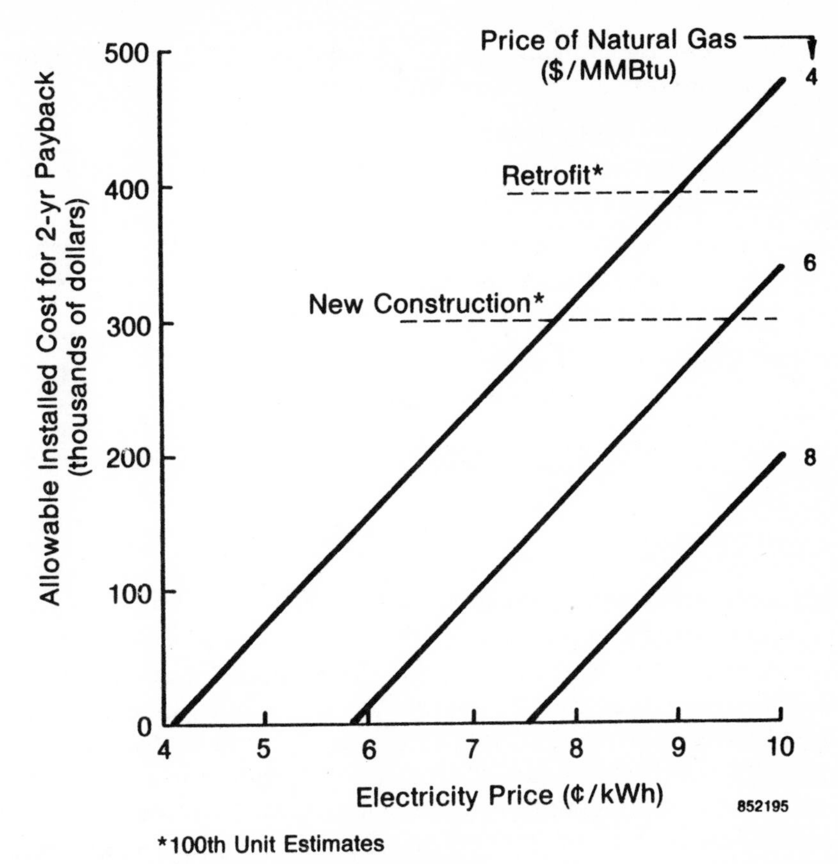

*100th Unit Estimates

Figure 10-7
ALLOWABLE INSTALLED COST, AS FUNCTION OF ENERGY PRICES

large enough to constitute a significant market for the system. Candidates include southern California, New York City, Long Island, and Connecticut, where there are favorable energy prices, large populations, and many light industrial facilities. These areas also contain many large commercial facilities, e.g., retail establishments and apartment complexes, that can employ this system.

FUTURE REQUIREMENTS

In order for the proposed system to appeal to a wider market, it is clear that user payback must be enhanced. Fortunately, many avenues exist for bringing this about, discussion of which is beyond this chapter's scope, but a brief listing is in order:

- Engine Improvements
  - optimization for natural gas, less reliance on off-the-shelf diesel hardware
  - adiabatic components (i.e., insulate and temperature-toughen engine so that more of the waste heat goes into exhaust, thereby improving chiller COP)
- Chiller Improvements
  - optimize $/ton for this size range
- Better Component Integration.

SUMMARY

Many types of cogeneration systems have been introduced and proven successful in recent years. Nonetheless, small-scale cogeneration still presents special problems. This chapter has described a system design that may be appropriate where other approaches are unattractive. The system uses reciprocating engines, and is fueled typically with natural gas. A typical configuration would be rated at 418 kW continuous-duty electrical output, but capacity can be varied over a range of 279 kW to 626 kW by making simple component substitutions.

Engine waste heat is used to drive an absorption chiller. The typical configuration would be rated at roughly 160 tons refrigeration capacity. This capacity can be achieved by using the chiller output directly, or by employing a refrigeration cascade to carry low-temperature refrigeration loads.

Economic projections indicate that presently the system is attractive only in selected geographic areas. Nonetheless, these areas have large populations of potential users and could provide a significant market. More importantly, potential system enchancements could significantly improve user economics, making the system attractive to a much larger market.

CHAPTER 11

Superposition, A Unique Cogeneration Opportunity

William L. Viar, P.E.

INTRODUCTION

The purpose of this chapter is to alert industrial managers to examine their plants for cogeneration opportunities that may have been overlooked, or missed by default. When first studied by the writer,[1] the particular "opportunity" was called Power Plant Superposition. Never was it dreamed at the time that the sound principle would someday be termed (a form of) cogeneration. The principle was valid because dual energy use was desirable then, as it is today. The timeless effects of superposition would conserve finite fuel resources. The basic differences in application of the concepts era-wise are: unlike 1953, fuel economics have changed dramatically; fuel availability is a concern in 1986.

SUPERPOSITION CONCEPT

As suggested, superposition is not a new idea. Perhaps the term needs clarification. It is the practice of building a higher pressure/ temperature power plant onto an existing industrial plant process heat and power load of consequence. Today, the name "topping plant" is often applied. The system objective is to continue to supply the industrial loads which heretofore had been supplied by relatively low pressure steam (say, 200 psig or less), and the franchised power utility in the area.

But with the superposed power plant, this base heat load permits the simultaneous or sequential production of high value mechanical shaft power at very favorable system heat rates, and at lowest practical energy cost.

The opportunities to cogenerate in this manner may be present on your plant for several reasons:

- When the steam system was originally built, both boiler fuels and power were relatively less expensive than now. There was little incentive to construct high pressure systems.
- The metallurgical limits and design knowledge were significantly different in years past. Materials technology has advanced remarkably since the advent of the electron microscope, for example.
- Power system designs and reliability have improved markedly as functions of need and innovation.

Even after high pressure and temperature steam generators and turbines evolved, there was no assurance that the need for moderate pressure steam would lead to properly cogenerated power. The installation of an 800 psig boiler did not necessarily prompt the installation of a companion turbogenerator. Too often, it appears, high pressure boilers—600 psig and up—were site-erected or delivered packaged only to operate at 200 psig or lower. The reasons for these kinds of equipment combinations may not be readily apparent, but could well have to do with an immediate industrial need for process steam, coupled with the cancellation by others of an order for a high pressure boiler. In such a rare instance, a bargain price for an HP boiler (although misapplied) may well have benefitted both parties.

Plant conditions that should lead to serious consideration of superposition will include:

- Process steam heat at 300 psig or less, and with mass loads of 50,000 pounds per hour or greater.
- An existing 600-800 psig, or higher, boiler operating at 200 psig or less.
- Old boilers must be replaced because of failures, and loss of reliability and dependability.
- The plant is being extended or expanded to increase production, or to add energy intensive product lines.
- Fuel conversion from oil/gas to coal is being planned.

- Relative cost of electric power is escalating at a greater pace than fuels for the forecastable future. Also, if the area utility is faced with imminent capacity expansion programs, superposition might help defer that investment.

The feasibility of this mode of power cogeneration is a direct function of the available enthalpy drop across the prospective turbine and the mass flow rate through it (as well as turbine quality). As an example, when 800 psig, 750°F steam is expanded through a high performance turbine at a rate of 50,000 pounds per hour, while exhausting to a 30 psig process load, the machine steam rate will be about 26 pounds of steam per horsepower-hour. Thus, for these conditions, over 1900 horsepower can be produced. If the actual steam flow were greater by a factor of four, then the power produced would be four times greater. If the Δ h available were double that described, then the power produced would be doubled.

A word of caution: if you contemplate building on your existing plant steam base, be sure that you have achieved all reasonable steam use economies prior to sizing the cogenerating equipment. Similarly, if a plant expansion is reasonably assured, increase the new turbine generator capacity to take full advantage of imminent new load. Do not oversize the equipment contemplating a substantial load increase 10 years hence.

The potential benefits of cogenerated power tend to be easier to calculate than are the estimates of costs to implement the opportunity. Therefore, it is essential that good cost information be obtained from equipment manufacturers or their vendors to achieve solid economic evaluations. Further, the benefits can be most accurately calculated and properly documented if the steam power system has been carefully simulated in a thermodynamic model of the facility. Incremental changes to be imposed on the system can then be fully tested and evaluated. Management information will be greatly improved.

INDUSTRIAL APPLICATIONS

The superposition ideas can be applied in more cases than one might imagine. Following are several examples known to the writer. Two of the plants have taken positive actions and have constructed significantly higher pressure/temperature boiler and turbine generator combinations over their adequately sized steam load bases. The other three have been advised of the existence of their opportunities, but no changes have been made to date. One of the two new cogenerators is a New England chemical plant that has converted from oil and gas to coal. The old boilers were removed, with the exception of one back-up unit, and the new stoker-fired boiler operates at 660 psig, 750°F. Rated load will be in the order of 150,000 pounds per hour. The back-pressure turbine generator set can produce 5,700 kilowatts. This system will sharply reduce the purchased power requirement for the site. The second superposition example was the Philip Morris Park 500 Complex near Chester, Virginia. That plant will be discussed further in the following section.

In the Philadelphia area, there is a chemical plant that has boilers designed for maximum working pressure of 500 psig, 600°F. Following the installation of the high pressure boilers, there has never been a process need to generate steam at rated conditions. Therefore, the steam has always been generated at a nominal 250 psig. About 2,500 horsepower are generated by auxiliary turbines as the system is now operated, but 2,000 kilowatts of new power could be produced under allowable steaming conditions. Required changes would include some boiler trim replacement, some external piping changes, and, of course, the installation of a new turbine generator. The chief advantage of this plant's opportunity is that no new boiler would be required.

A western oil refinery, subjected to severe weather conditions, requires an average of 200,000 pounds per hour of steam for its many boil-up and separation processes, along with major nonprocess heating loads. Steam is generated and distributed refinery-wide at a nominal 150 psig. If a 1,250 psig, 900°F boiler and back-pressure turbine generator set were superimposed on this load base, 9,500 kilowatts of new power could be generated. This exceeds present purchased power requirements by 1.5 megawatts. The economics will be tougher here than in the previous example because of the need to buy the boiler as well as the turbine generator. However, because of proximity to vast reserves of western coals, a fuel conversion for steam and power production would increase the economic incentive to superpose.

A chemical plant in Ontario, Canada, presently purchases 7.8

megawatts from Ontario Hydro, and generates over 400,000 pounds per hour of steam to drive its processes. Steam conditions are 425 psig, 650°F. Natural gas is the normal fuel, with Bunker C oil as backup. There are two possible changes of consequence at this facility that could improve steam utilization. First, in the near term, two of the active boilers are rated to operate at 775 psig, 750°F. If these two boilers, which normally carry 82 percent of the total plant load, were simply upgraded to operate as originally designed, and a new back-pressure turbine generator were installed, new power in the amount of 4.5 megawatts could be produced. Again, this involves investment in boiler trim and some external piping, but no new boiler. Therefore, this superposition opportunity appears to have high merit.

A second possible change at this plant location is a switch from gas/oil to coal. The plant has been advised that should this basic retrofit be funded, the new coal-fired unit should definitely go in at a high pressure and temperature, coupled with a back-pressure turbine generator.

"New" power generated in the manner described above is not cost free, but it does often represent the lowest cost power that you can produce. The heat converted to work is chargeable to the power producing turbine, but the delivery of its exhaust steam to valid process and service loads results in highly favorable system incremental heat rates.

PLANT THERMODYNAMIC MODEL

The Philip Morris Park 500 major cogeneration plant is commented on here. A computer model was developed from a steam system schematic diagram. The system was simulated as depicted. Computer software known as MESA[2] was used to calculate the detailed mass and energy balance indicated by the schematic flow sheet.

With this model, the system can be tested for operational strategies that involve process load changes, system component isolation and shutdown, the effects of part-load operation of the turbogenerators, or other changes. For example, should the next kilowatt load be applied to the double extraction, back-pressure unit or to the pure back-pressure unit? Rules of thumb or intuition will not answer that type question adequately as to direction and magnitude of change.

The highest pressure unit was designed to generate 1,650 psia steam at 950°F. There are nominal pressure drops across the superheater and the lead line to the turbine, so that the throttle conditions are 1535 psia, 950°F. Mass flow rate through that system is 204,400 pounds per hour. Regenerative feedwater heaters are supplied steam from extraction (bleed) points at 520 psia and 205 psia. The deaerator is supplied steam by the turbine exhaust at 45 psia. These throttle, extraction, and exhaust conditons are indicated in Figure 11-1 on the curve for Unit A. This curve approximates the steam condition line for expansion through this major turbine. See Figure 11-2 for the computer model diagram.

The 850 psia boiler was described as coal-fired with oil backup that had originally operated at 200 psig. After the pressure drops across the superheater and the lead line, throttle steam was shown at 685 psia, 750°F. This unit exhausts at 205 psia, in parallel with the large machine second extraction. With a flow of 196,000 pounds per hour through the turbine, generator output is reported at 5.7 megawatts. There is some desuperheating indicated on flow streams from both units. In Figure 11-1, the steam condition line is shown as the Unit B curve. There is one stage of regenerative feedwater heating for this 850 psia steam generator.

Plant steam and power loads established the need for this cogeneration plant. Steam loads supplied from the 205 psia header total 280,000 pounds per hour, including regenerative feedwater heater steam. The 45 psia header supplies loads in the amount of 117,400 pounds per hour, counting deaerator steam. Condensate recovery at this facility is 82 percent of the total feedwater.

It is not known at this time what investment Philip Morris committed to this energy effective complex. On the basis of system analysis, several tentative conclusions can be drawn concerning energy operating costs and benefits:

- To satisfy process and service steam requirements, and to generate 18.2 megawatts, boiler fuel in the amount of 472.1 x 10^6 Btu per hour is required.

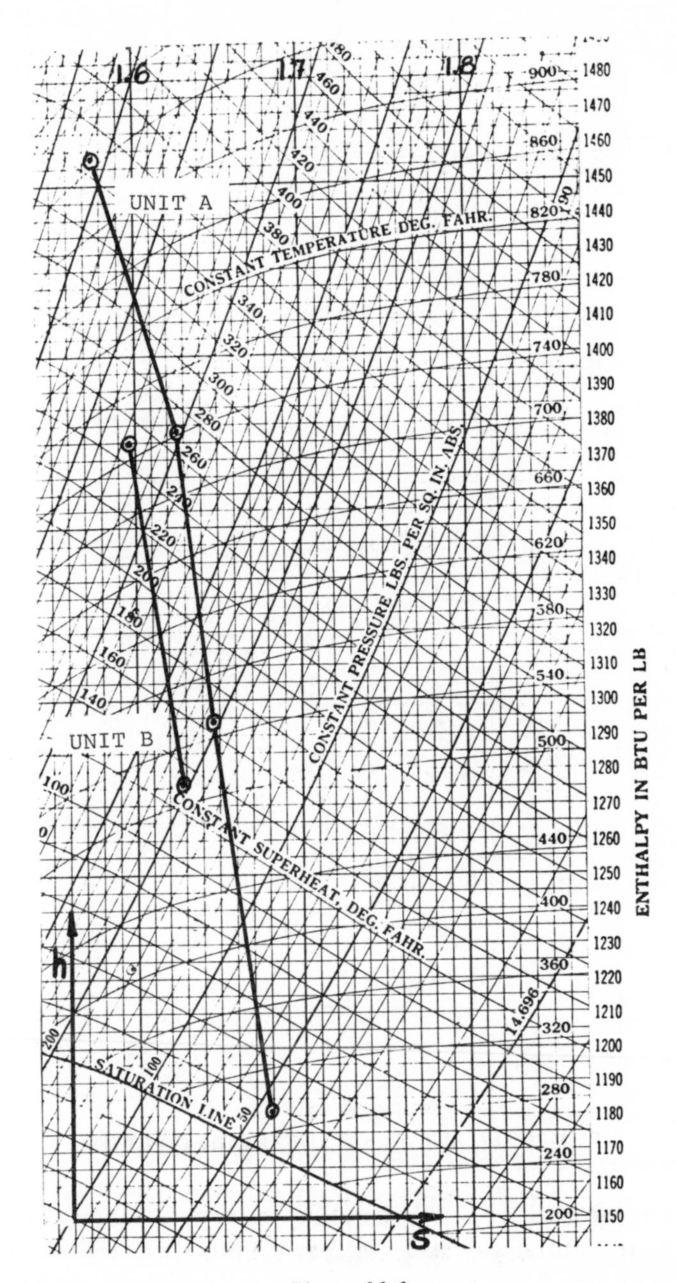

Figure 11-1
EXCERPT FROM COMBUSTION ENGINEERING MOLLIER DIAGRAM
Approximate steam expansions through new turbines. Unit A, 12.5 MW, double extraction, back-pressure. Unit B, 5.7 MW back-pressure.

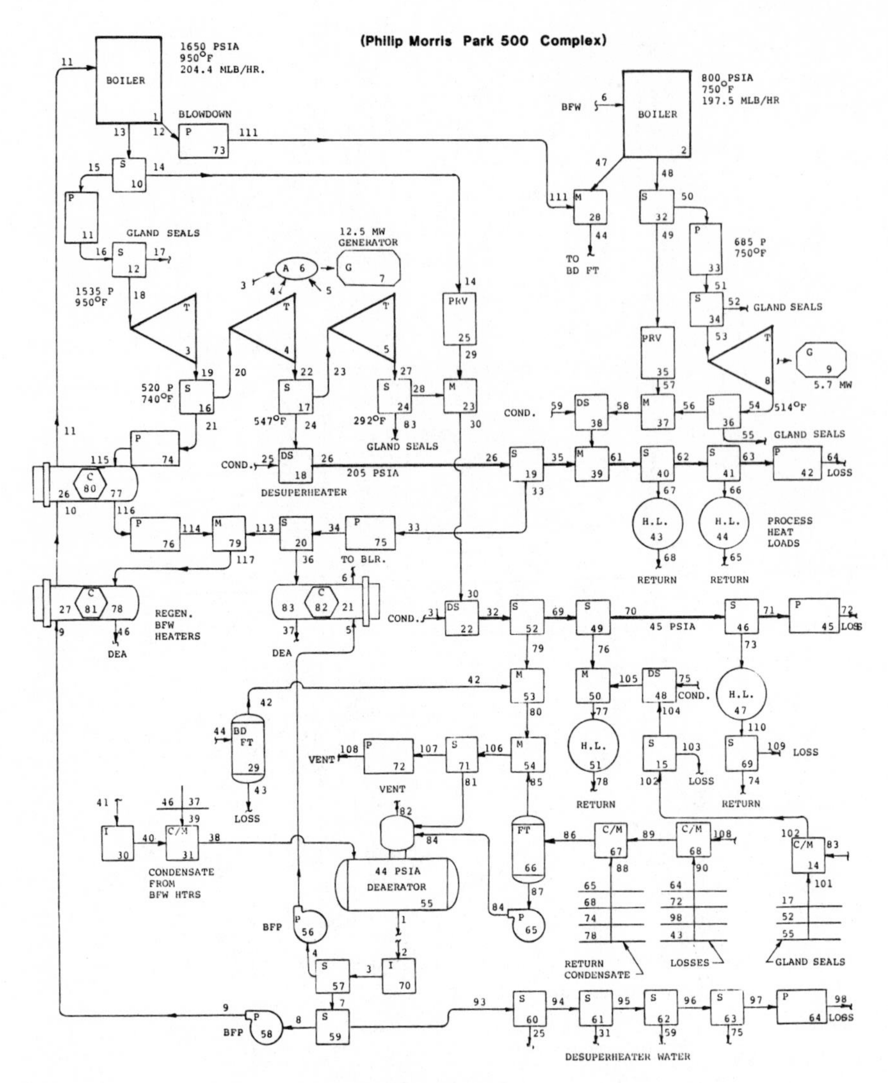

Figure 11-2
COMPUTER MODEL OF SUPERPOSED PLANT

- On the basis of 8000 operating hours per year, and if the fuel were natural gas at $5.00 per million Btu, the annual cost would be $18.88 million.

- If heavy oil at $4.40 per million Btu were used, the annual fuel cost would be $16.62 million.

- With coal for fuel, and if it were $1.80 per million Btu, the annual cost is about $6.80 million.
- The apparent fuel saving per year for coal over oil is $9.82 million.
- If the factory were to use all the power cogenerated, and avoided its purchase at a rate of 3.66¢ per kilowatt-hour, the annual savings would be $5.33 million. There could be a favorable buy/sell arrangement with the utility that would result in an additional moderate saving for the owner.
- The rough energy saving estimate for this cogenerating facility is $15.15 million per year, relative to oil-fired low-pressure steam.

Admittedly, assumptions have been made to appraise this cogenerating plant. The financial benefit to the owner may well be less than, or greater than, the suggested numbers. However, the exercise demonstrates the competence of Philip Morris Management and its Consultants, United Engineering & Constructors, Inc., of Boston. Further, it exemplifies the validity of the expressed need for review of your plant for similar superposition potential.

CONCLUSIONS

Superposing or topping a high pressure/temperature power plant over an appropriate heat and power base load is thermodynamically feasible for industrial plants. Often, there will be acceptable financial and economic merit. Industrial managers should examine their opportunities of this type and evaluate them. The potential benefits can be large; the costs of implementation may be within your limits for return on investment. The unusual cogenerating technique should not be laid aside *a priori*.

References

[1] F.T. Morse, P.E., *Power Plant Engineering,* D. Van Nostrand, 1953, 3rd Edition.
[2] The MESA Company, 22 Golden Shadow Circle, The Woodlands, Texas 77381.

CHAPTER 12

Design and Economic Evaluation Of Thermionic Cogeneration In A Chlorine-Caustic Plant

Gabor Miskolczy, Dean Morgan & Roger Turner

INTRODUCTION

Studies show that it is feasible to equip a chlorine caustic plant with thermionic cogeneration. Thermionic combustors replace the existing burners of the boilers used to raise steam for the evaporators, and are capable of generating approximately 2.6 MW of dc power. This satisfies roughly 5 percent of the chlorine cells' power demand. More thermionic power could be generated, and excess steam would be produced which could be sold or used elsewhere in the plant. A typical plant was defined based on a survey of U.S. chlorine plants. This plant produces 470 U.S. tons of chlorine per day, with four cell rooms. Each cell room is one electrical circuit and requires a dc supply of 185 V and 70,000 A. Total dc power consumption is nearly 13 MW. The steam for the evaporators is raised in four boilers with a total installed capacity of 320,000 lb of steam per hour.

The study shows that the estimated cost of thermionic cogeneration installed in the typical chlorine-caustic plant is $1600 per kW.

THERMIONIC COGENERATION SYSTEM DESCRIPTION

A block diagram of the cogeneration system is shown in Figure 12-1. This figure shows the thermionic combustor interposed between the burners and the boilers. The electrical output of the thermionic combustor is fed to the cell room through a bus system and protective switches, as shown schematically in Figure 12-2. Hot gases generated by the burner heat the thermionic converters and then pass to the existing boiler, where steam is raised for the evaporators.

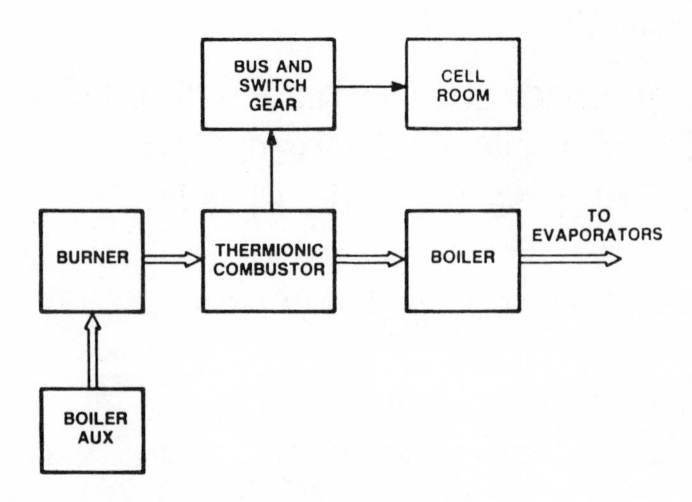

Figure 12-1
BLOCK DIAGRAM OF COGENERATION SYSTEM

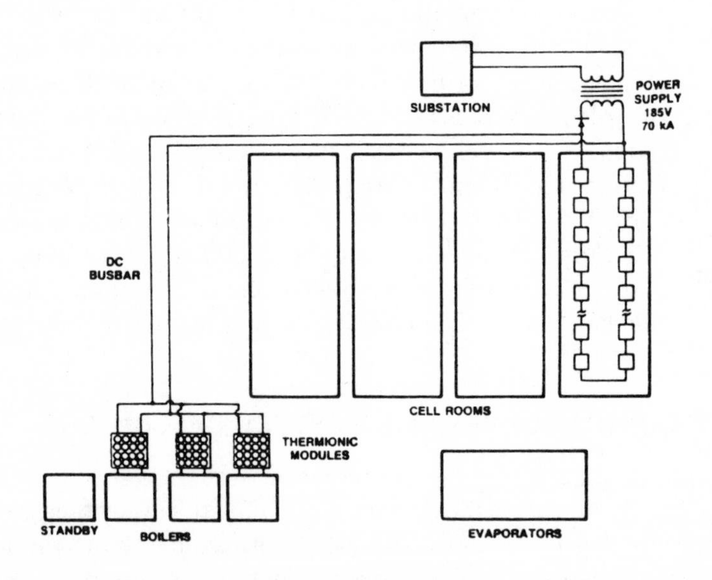

Figure 12-2
SCHEMATIC LAYOUT OF
PLANT WITH THERMIONIC COGENERATION

The typical plant was defined from the characteristics of the various plants in an earlier survey.[1] To protect the confidentiality of the sources of the plant survey, a typical plant, rather than a specific actual plant, was used in this design.

The typical plant consists of four-cell rooms. Each cell room has one cell circuit of 50 cells requiring 185 V and 70,000 A dc. The power for the four-cell rooms is supplied by a common substation. Four boilers, one of which is on standby, are located approximately 210 m (700 ft) from the cell rooms. The distance is measured along the existing pipe racks.

The relative electrical and thermal requirements of a diaphragm cell chlorine plant are such that generation of all the direct current required for the cell rooms by cogeneration would produce an excess of process steam. In some instances, such as a chlorine plant which is part of a larger complex, the excess steam could be used elsewhere in the plant. For the present study, it was assumed that the excess steam could not be used or sold. Thus, the amount of direct cogeneration was set by the steam generation.

It was found that if three of the four boilers, each of which has a capacity of 40,000 to 120,000 lb/hr of steam, are equipped for thermionic cogeneration and produce the steam required by the plant, a total of 2.5 MW can be generated. Each cell room requires about 13 MW. Therefore, the thermionic cogeneration system could supply nearly 20 percent of the power required for one cell room. The thermionic power would be connected in parallel with the existing rectifier and would supply the full load voltage of 185 V. The amount of electric power that can be generated for each Btu fired in the burners has been calculated with the process gas temperature as a variable. It was shown[2] that the maximum thermionic power produced is 18 kW per million Btu fired per hour. All combustors are similar but progressively larger in size to match the thermal output of the boilers.

A longitudinal cross-section view of the combustor is shown in Figure 12-3.

The combustor is an annular type. The combustion volume is surrounded by a cellular ceramic radiant body. The combustion gases pass through the cellular ceramic and heat it to a high temperature to enhance the radiant heat transfer to the thermionic emitters which surround the cellular ceramic. The collector cooling air passes

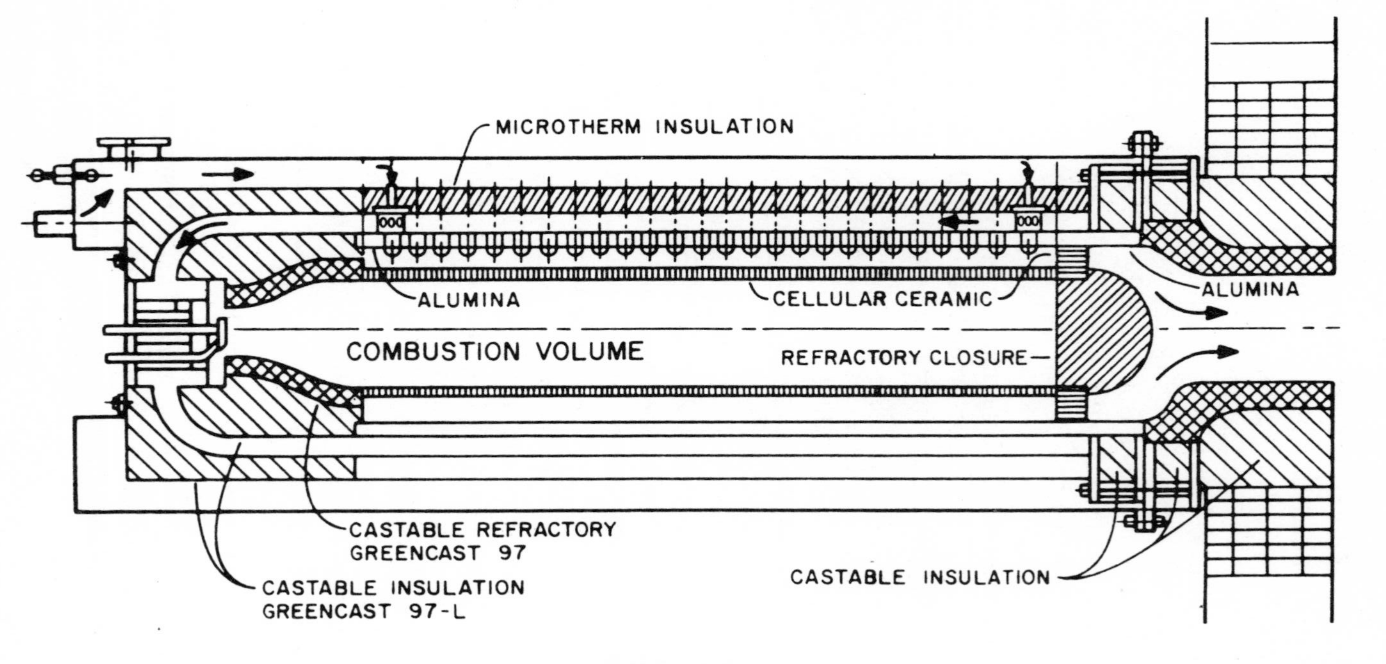

Figure 12-3.
THERMIONIC COMBUSTOR CROSS SECTION

through the collector heat exchangers and exits to the burner as preheated combustion air. The hot combustion gases pass by the thermionic converters and exit to the boiler through a refractory lined tunnel.

The thermionic combustor consists of: 1) the array of thermionic converters supported on a suitably insulated structure, 2) the burner, and 3) the mounting to the boiler burner wall. Five thermionic combustors are mounted on the front wall of each boiler. Each combustor is suspended by rod hangers from the main supporting beam. There are three sizes of types of combustors, each tailored for one boiler size in the plant as shown below.

Combustor Type	Boiler Steam Output lb/hr
I	40,000
II	80,000
III	120,000

THERMIONIC CONVERTER ARRAY

The thermionic converter used in the design of the array is a 2-inch nominal diameter emitter with a torispherical hot shell consisting of a chemically vapor deposited (CVD) tungsten emitter and a CVD silicon carbide protective surface.[3] The design has been adapted[4] for large output leads to minimize the voltage drop between converters connected in arrays.

Connection of converters in a series-parallel array to achieve the high voltages and high currents normally required to ensure easy power transmission and good reliability has long been recognized as necessary in most designs which use large numbers of thermionic converters. The three combustor designs projected for use with the boilers in the chlor-alkali plant would each use a slightly different arrangement of converters.

The arrays are connected in series around the circumference of the combustors. Each converter is mounted, electrically insulated from the support structure, by an integral crimp-on flange.

SYSTEM PERFORMANCE

The thermionic converter used in the system was assumed to have the following characteristics:

Emitter temperature	1750 K
Collector Temperature	900 K
Incident Thermal Flux	39.6 W/cm^2
Thermal Efficiency at the Terminals	13.65%

The thermal requirements of the plant were assumed to be satisfied by three of the four boilers. The fourth was considered a standby. The capacities of the three boilers were 40,000, 80,000 and 114,000 pound steam per hour. The thermal input of each of the boilers was subdivided into five thermionic combustors. Each boiler was to be equipped with different sizes of thermionic combustors so that the thermal capacity of five of each type combustor would match that of the boiler. The number of diodes or converters in each combustor was based on matching this thermal output.

The performance of each of the types of combustors is shown in Table 12-1. The electrical characteristics of each type of combustor depend on the number and arrangement of the converters in each type. Type I has approximately 822 converters arranged in series-parallel strings; Type II has 1628 converters arranged in 8 series strings each of which has 7 or 8 converters in parallel. Type III has 2457 arranged in 7 series strings of 52 with 7 or 8 converters in parallel. Each arrangement produces an operating array output of 192 V dc with currents that are approximately 473 A, 947 A, and 1457 A respectively. The method chosen for conducting the start-up sequence is to disconnect the load and temporarily attach a start-up circuit which ignites the converters in the array by a capacitor discharge. This arrangement eliminates power transients into the load and limits the amount of energy dissipated in the ignition transient to a safe level.

The Type I, II, and III combustors have similar I-V curves, as shown in Figure 12-4. The ignited portion of the I-V characteristics intersects the unignited mode at approximately 270 V, and the short-circuit current is about 1629 A in Type I. If, starting from the

Table 12-1
POWER OUTPUT OF THERMIONIC COMBUSTORS

Combustor Design	Boiler Steam Output (lb/hr)	Numbers of Combustors	Power/ Combustor (Watts)	Total Power for 5 Combustors (MWe)
Type I	40,000	5	86,563	0.433
Type II	80,000	5	173,126	0.866
Type III	120,000	5	259,689	1.298
			Total	2.597

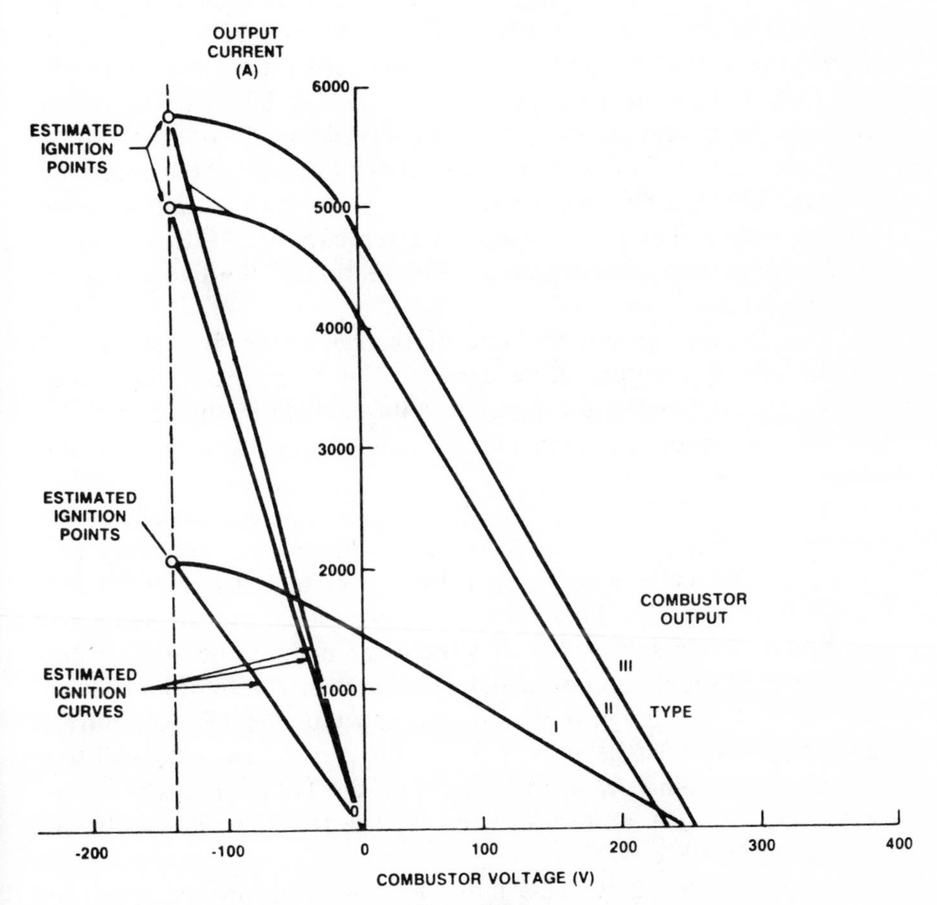

Figure 12-4
I-V CURVES FOR TYPE I, II, and III COMBUSTORS

unignited mode, a voltage is impressed on the array to drive the converters out of the power quadrant, it is expected that a "staircased" transient will occur as various converters in the array ignite. The current grows upward to a point eventually reaching the ignited mode characteristic at a total of close to 220 A at 160 V. At this point, the voltage would be divided to produce ½ V per cell across each of the converters in the string.

The performance analysis of the system includes an analysis of the operation of the TEC combustors as connected to the cell system, and an analysis of the operation on the protective equipment included in the bus system.

The operating and control characteristics of a chlor-alkali cell system are demonstrated graphically in Figure 12-5. The figure shows the operating and control characteristics of the "typical" cell room before the connection of the TEC combustors. The lowest curve is the operating curve for the chlor-alkali cell room circuit; it indicates the current drawn by the cells as a function of the impressed voltage. The rectifier operating curve is also shown; it indicates the voltage setting required to supply a given current. At design conditions, the system operates at the intersection of the two operating lines (Point B).

The chlorine production rate of the cell system is proportional to the cell system operating current. The production rate is controlled by controlling the rectifier voltage which in turn moves the rectifier operation curve by changing its voltage intercept with the slope constant.

The operating lines for the TEC combustors are straight in the region of interest, as also shown. The thermionic operating line includes the voltage drop through the bus system. The rectifier and combustor operating lines have been combined and are indicated as shown. The results show that the combined rectifier-TEC installation would supply approximately 73 kA (Point C). In order to keep the plant at the design point, the rectifier plus the TEC combustor operating line must be moved downwards to intersect the cell line at the design point, A. At this condition, the TEC combustor will be supplying 13.6 kA at 185 V, Point A. The rectifier will supply the remaining 57.4 kA.

Inspection of Figure 12-5 indicates that during normal operation near the design point, the TEC combustion output will ride on the

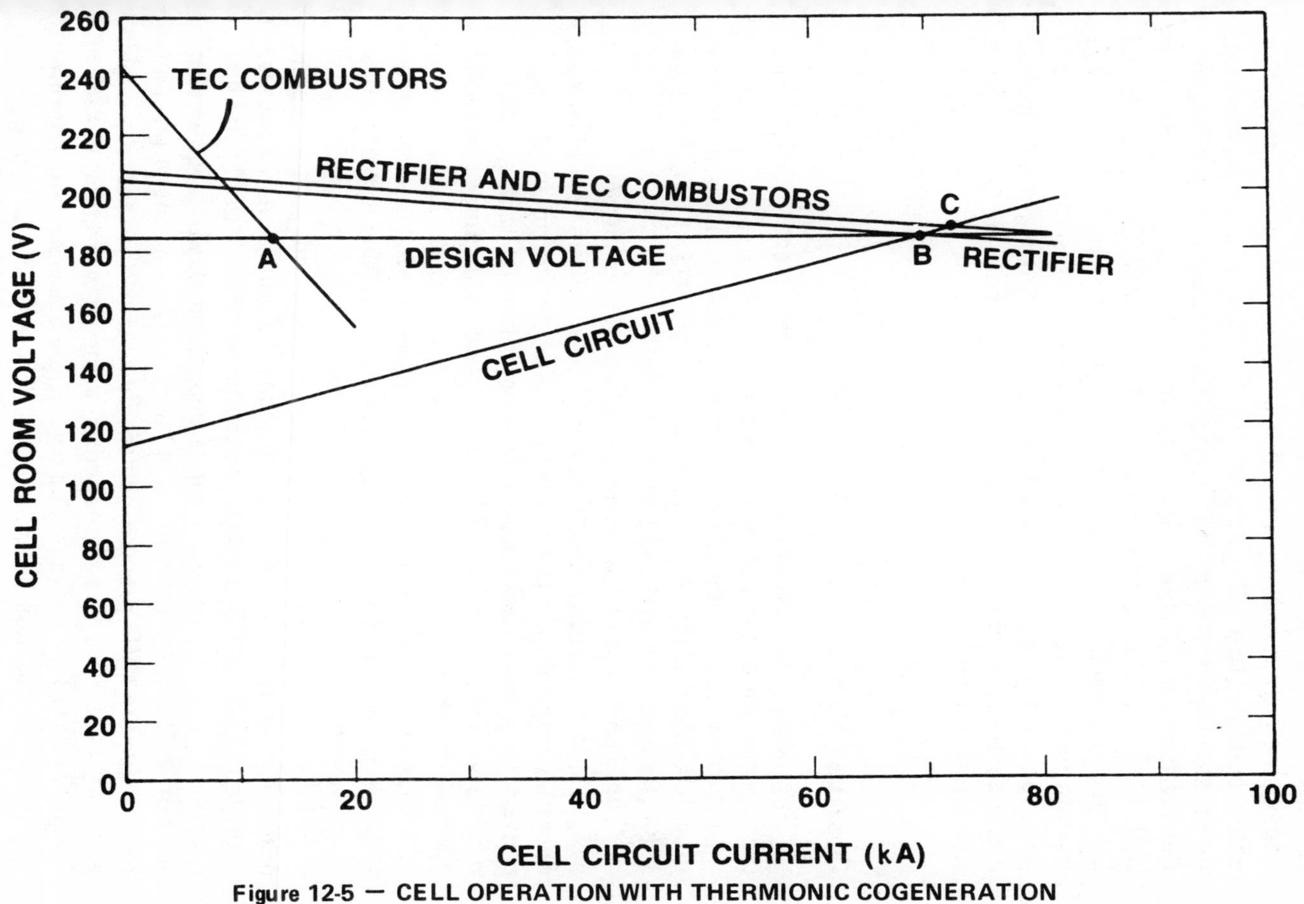

Figure 12-5 — CELL OPERATION WITH THERMIONIC COGENERATION

rectifier control. As the rectifier output is decreased, the TEC combustor output will increase to the capacity of the system. This action is similar to a cell system when parallel power supplies are used.

In summary, with the TEC combustors installed in parallel with the rectifiers, the system can be controlled around the design point with the normal rectifier controls. As the operating rate is decreased by "tapping back" the rectifier, the TEC combustor output will increase until a point is reached where its bus system capability is reached. At this point, the bus system protective equipment will disconnect the combustors from the cell circuit and the rectifiers will supply the entire load current.

COST ESTIMATES

The cost of equipping the typical chlorine plant with a thermionic cogeneration system has been estimated.

Converter fabrication costs were assessed several times in previous programs. Specifically, costs were assessed in 1975, 1979, 1980 and 1984 (References 5, 6, 7, and 4).

These cost studies have generally assumed then-current converter designs and fabrication practices could be scaled up (either in size, operating capabilities, or production numbers) to practical system converters without adversely affecting either fabrication cost or reliability. They have been directed toward estimating a converter purchase price per kilowatt under that assumption. Thus, the cost estimates should be regarded as a projection of what a more mature technology should achieve, rather than as a strict measure of present thermionic costs.

The converter costs in Reference 4 are based on the present design of the two-inch diamater thermionic converter which uses a composite hot shell-emitter structure (silicon carbide-graphite-tungsten) made by chemical vapor deposition (CVD). The cost of the hot shell-emitter structure was based on the amount of reactants needed for the CVD process, current quotations for these materials, and projected labor and capital costs assuming a production rate of 200,000 converters per year. The cost of the balance of the converter was based on the data in Reference 6. The prices in this study were increased to account for inflation. Variations in component sizes were adjusted for by assuming the cost to vary with the square of the converter diameter.

The converter cost assumptions were reexamined for this chapter. It was found that the raw material costs were realistic. However, a consensus of the individuals reviewing the labor costs associated with the converter construction was that the earlier estimates were valid in very high volume production after extensive manufacturing experience and quite large capital investment. For an initial application such as considered here, labor costs associated with chemical vapor deposition, bonding and processing of the converter were increased by a factor of ten.

It is evident that the order of magnitude increase in labor cost to manufacture the converters translates into an increase in converter production cost from 199 to 554 $/kW for a nominal 2-inch diameter converter. This revised converter cost is much more conservative than that given in Reference 4.

On reviewing the system costs in Reference 4, it became evident that the A&E cost estimates assumed field construction of the thermionic combustor. Thus the costs of erecting the converter support structure and insulating the combustor were based on site construction. However, the size of the thermionic combustors allows factory assembly with significant savings due to the lower cost of shop labor and the greater efficiency of factory operations.

Examination of the other system costs (i.e., burner, burner support, boiler modifications busbar, ignition circuit and electrical switchgear) in Reference 4 showed that they were realistic. The material costs associated with the non-converter components of the thermionic generating system given in Reference 4, appear to be valid.

The installed thermionic generation cost for retrofitting a chloride caustic plant was revised based on the foregoing considerations, namely—

1. A converter cost or $554 per kilowatt (corresponding to an order of magnitude increase in labor costs)

2. No revision in material costs

3. No revision in the cost estimates for electrical switchgear and busbars

4. No revision in the cost estimate for ignition circuit

5. Reduced thermionic combustor construction cost (based on factory assembly rather than field erection)

6. Engineering cost as 15 percent of the cost of installing the burner modules only.

The results of this calculation are shown in Figure 12-6, which is a pie chart with the percentage of cost for each element. The revised total cost for retrofitting thermionic generation into the chloride-caustic soda plant is $4,162,000, corresponding to a specific cost of $1603/kW. Although the revised cost estimate is 5.8 percent higher than that in Reference 4, one concludes that these more realistic installed thermionic generation costs are economically competitive.

The effect of converter performance and size on converter cost was evaluated. The converter power density was varied from 5 to 6 W/cm^2. The converter size was varied from a nominal 1-inch diameter to 6 inches. The results show that the lowest converter costs are obtained with converter diameters of about 4 inches. For these calculations the dimensions in the emitter sleeve were changed so as to result in the same percent of total power lost in the sleeve for the cases studied.

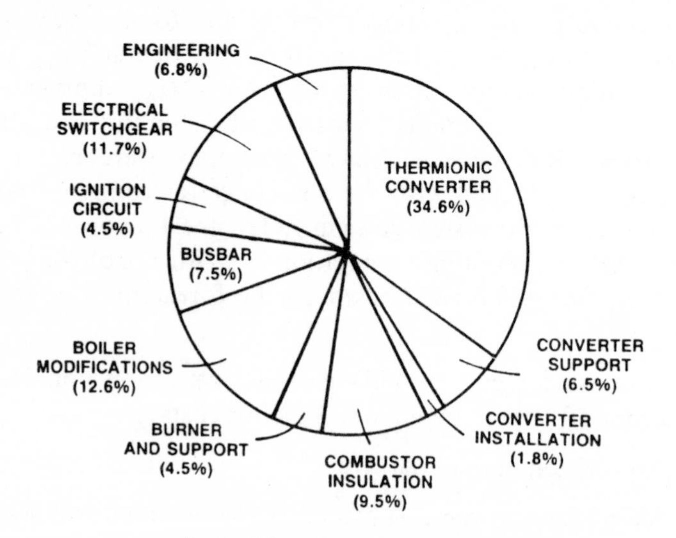

Revised Cost Per Kilowatt ($1603/kW)

Figure 12-6
THERMIONIC COGENERATION SYSTEM COSTS

The lowest converter cost is calculated for a nominal 4-inch diameter converter with a cost of $400/kW vs the $554/kW assumed in the cost study. It is seen that the optimum size of the converter is increased from 2 to 3 inches in the earlier study to 4 inches for this chapter. The system costs here have been calculated for the 2-inch converter, since the 2-inch converter would be used in an initial installation, and the combustors were designed to accept the 2-inch converters.

ACKNOWLEDGEMENTS

The collaboration of A.E. Margulies and R. Deluca of Stone and Webster Engineering Corporation and Reay Dick of Rasor Associates, Inc. is acknowledged.

References

[1] R.A. Turner, "Chloride-Caustic Soda Plant Survey and Data Base," *Topical Report* TE4333-6-84, January 1984.

[2] G. Miskolczy and D. Lieb, "Cogeneration Using a Thermionic Combustor," *Proc. 17th IECEC,* Aug. 1982.

[3] Advanced Thermionic Technology Program, Summary Report TE4258-5-84, DOE/ET/11292, Oct. 1984.

[4] Final Report: "Thermionic Energy Conversion Design for Chlorine-Caustic Soda Application," Contract DE-AC02-83CE40635, Rpt. No. DOE-CE-40635-2 (1984).

[5] Anon, "Phase One Report, Thermionic Conversion Topping System Study," Thermo Electron Report RE4207-29-76, Dec. 1975.

[6] G. LaRue, G. Miskolczy, "Manufacturing Cost of Flame Heated Thermionic Converters," Thermo Electron Report TE4258-163-79, April 1979.

[7] G. Miskolczy, C.C. Wang, B.T. Lovell, J. McCrank, "Topping of a Combined Gas and Steam Turbine Powerplant Using a TAM Combustor," Thermo Electron Report No. 4258-167-81, March 1981.

PART IV

INDUSTRIAL COGENERATION: CASE STUDIES

CHAPTER 13

Gas Turbine Cogeneration Plant For the Dade County Government Center

Roger W. Michalowski and Michael K. Malloy

INTRODUCTION

Dade County, Florida, has been implementing a phased development plan to consolidate its administrative activities in Miami. Desiring to revitalize the downtown area and provide an attractive working environment for their employees, they have complemented administrative buildings with a cultural center. A "Metrorail Transportation System" and the "People Mover," a track-mounted, rubber-tired vehicle system under construction, link the Government Center with downtown Miami and outlying areas to alleviate impending traffic congestion.

This well-conceived complex also houses a unique energy system for the supply of electricity, air conditioning, and domestic hot water. Locating such a powerplant in downtown Miami presents significant engineering, environmental, and construction scheduling challenges.

Selecting the appropriate system took many preliminary design iterations. Energy services are supplied to the five prominent complex buildings: the Dade County Administration Building, the Revitco Building (housing the GSA), the Dade County Courthouse, the Central Support Building, and the Cultural Center. The primary thermal demand of air conditioning from these buildings varies substantially over both daily and seasonal profiles. The cogeneration system to be provided incorporates the flexibility necessary to optimize fuel utilization efficiency while serving such variable thermal demands of the complex. Sized to take into consideration future Dade County expansion plans, the system will efficiently produce additional electricity when chilled water demands are low.

The cogeneration plant consists of a Rolls-Royce gas turbine-generator set and a waste-heat recovery system which recovers waste heat from the gas turbine exhaust. The waste-heat recovery system consists of a Zurn dual-pressure, heat recovery boiler, a Thermo Electron dual-pressure, extraction/condensing steam turbine generator set, and four Trane absorption chillers. Steam is produced in the waste-heat recovery boiler at two pressure levels. The high-pressure steam is directed to the steam turbine for the production of electricity. The low-pressure steam is directed either to the steam turbine or the four Trane absorption chillers for the production of chilled water. Condensate from the absorption chillers is directed to a heat exchanger for the production of domestic hot water.

Figure 13-1 shows the cogeneration topping cycle flow diagram for the facility. The specific useful energy streams are as follows:

- *Power Generation*

  –A Rolls-Royce gas turbine-generator set with a continuous site rating of 22,000 kWe, firing natural gas. The turbine is coupled to a Brush two-pole, cylindrical, brushless ac generator.

  –A Thermo Electron multistage, dual-pressure, extraction/ condensing steam turbine generator set with a maximum continuous rating of 10,000 kWe.

- *Air Conditioning*

  Four Trane absorption chillers utilizing low-pressure steam have a total rated output of 5200 tons of refrigeration. The chilled water leaving the absorption chillers will be piped to the chilled water distribution pumps, which are designed to discharge at a maximum rate of 9200 gpm through the chilled water distribution system which supplies the downtown Government Center.

- *Hot Water*

  Condensate from the absorption chiller is delivered to the hot water heat exchanger which is capable of producing up to 300 gpm of hot water for domestic use.

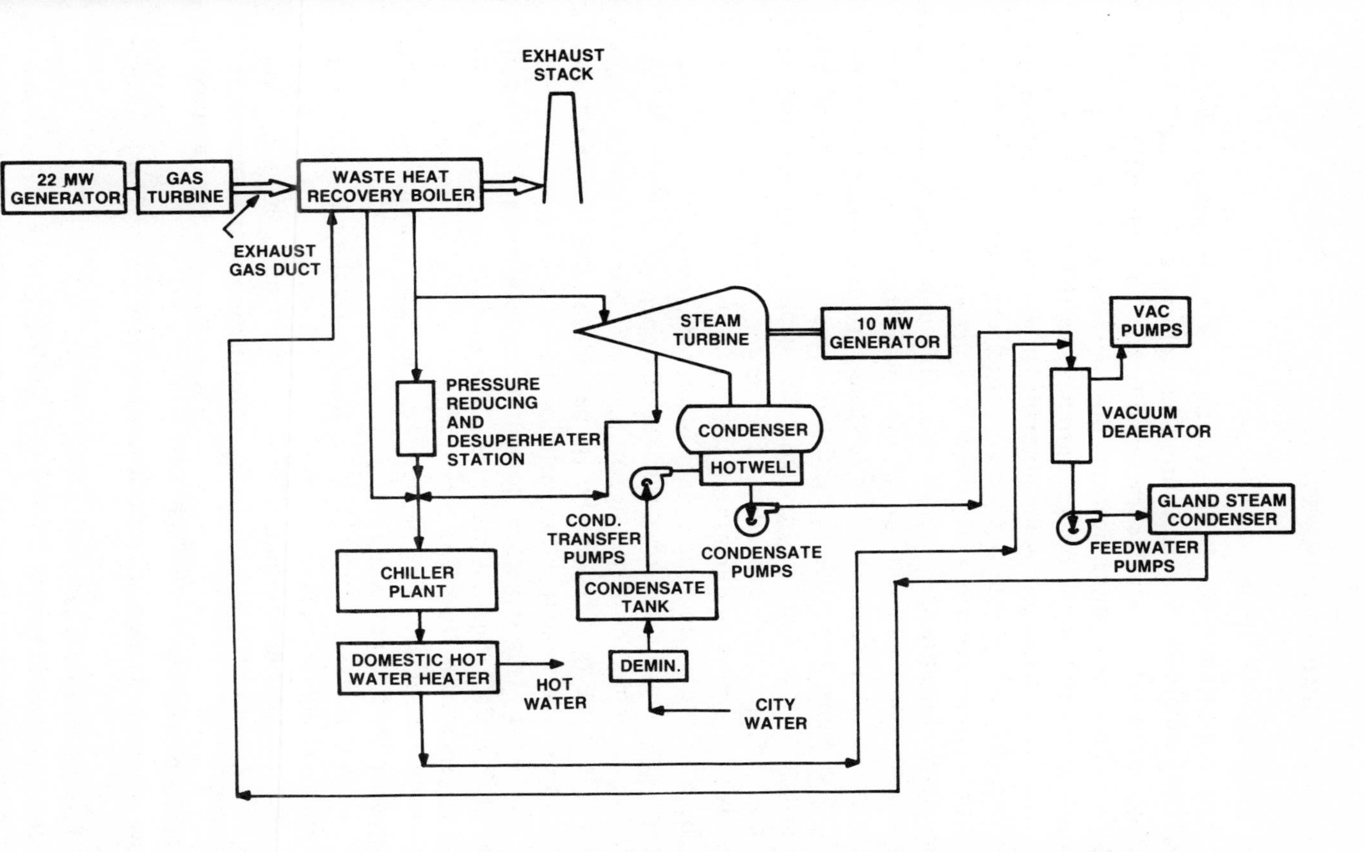

Figure 13-1 — DADE COUNTY COGENERATION PLANT TOPPING CYCLE FLOW DIAGRAM

MAJOR EQUIPMENT COMPONENTS

Gas Turbine-Generator Set

The Rolls-Royce gas turbine-generator set is equipped with dual fuel-firing capability, normally firing natural gas, with No. 2 oil used during abnormal or emergency conditions. The natural gas fuel supply is boosted from service pressure to 330 psig for firing in the gas generator section of the gas turbine. Two 100-percent capacity gas booster compressors, one normally in operation and one on standby, are provided for gas compressions. In the gas generator, the air is compressed in a five-stage, low-pressure compressor and a seven-stage, high-pressure compressor. This compressed air is used for combustion of natural gas in the combustion chamber, and is then expanded through two single-stage turbines driving the compressor sections. The pressurized high-temperature exhaust from the gas generator then expands through a three-stage, free power turbine driving the electrical generator. The power turbine is connected directly to the generator and rotates at 3600 rpm.

Because the cogeneration plant is located close to the coastline, the gas turbine air intake design is equipped with a three-stage filtration system. The first stage utilizes self-cleaning inertial separators. Bleed air (approximately 10 percent of the total air flow) carries away the separated dirt particles. The second stage is a coalescer pad filter. This stage acts to remove moisture that has passed through the first stage inertial filter. The third stage uses high-efficiency, replaceable, bag-type filters. This stage is used primarily for the extraction of very fine particles, including salt, which were not filtered by the two preceding stages.

The noise emission from the air intake structure is reduced to environmentally acceptable levels by the installation of splitters into the vertical intake duct. The splitters attenuate high and low frequency noise and are of frame construction and designed for long life. They are padded with suitable sound-absorbent material, such as mineral wool, covered with close-weave sheeting, and the entire assembly is clad with perforated galvanized steel or stainless steel.

The gas turbine's NO_X emissions are guaranteed to be approximately 75 ppm (adjusted for 15 percent oxygen on a dry basis), without water or steam injection. The allowable emissions established by the Standards of Performance for New Stationary Sources pertaining to gas turbines of this efficiency range is approximately 90 ppm NO_X.

Waste-Heat Recovery Boiler

The exhaust gas from the gas turbine which is approximately 1010°F at full load is ducted to an unfired dual-pressure, natural-circulation, waste-heat recovery boiler. The dual-pressure design is employed to maximize useful steam production. The boiler consists of two sections, high pressure and low pressure, and will generate approximately 95,000 lb/hr of steam at 625 psig, 810°F and 22,000 lb/hr of steam at 20 psig, saturated conditions at full load. Exhaust gas exiting the boiler is ducted to a dual-wall steel exhaust stack.

The feedwater and steam flow through the boiler is shown in Figure 13-2. The boiler feedwater is pumped from the deaerator to the LP economizer and flows to the LP drum. Saturated water from the LP drum flows through the LP boiler. Steam from the LP drum is supplied to the LP steam system. Saturated water from the LP drum is used as feedwater for the HP boiler section. The saturated steam from the HP drum is superheated to 810°F at the superheater outlet. The boiler water quality is monitored and controlled through the combined use of boiler blowdown and chemical addition.

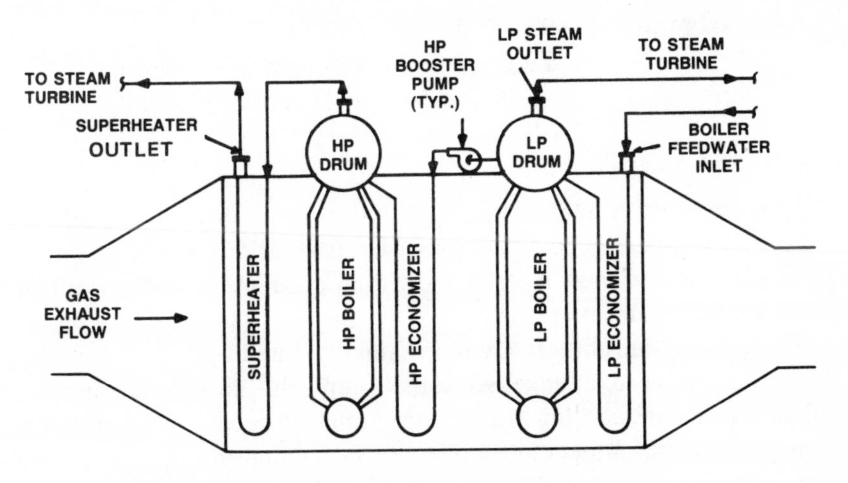

Figure 13-2
DUAL-PRESSURE, WASTE-HEAT RECOVERY BOILER

Steam Turbine Generator Set

All the steam from the boiler flows to the steam turbine under normal operating conditions. The steam is a dual-pressure, multi-stage impulse design with a throttle inlet designed for 625 psig and an induction inlet for 20 psig. A sliding gate pneumatic valve is mounted prior to the induction port for start-up and emergency conditions. The steam turbine also has a single automatic extraction controlled at 20 psig to supply low-pressure steam for use in the absorption chillers. A rotary disc valve controls the back-end turbine flow in response to the chiller demand. The last stages of the turbine exhaust steam to a condenser optimized for the average condition of 3½ in. HgA. The turbine is coupled to the generator through a reduction gear. The turbine design speed is 6000 rpm and the generator rotates at 1800 rpm.

The following lists the steam turbine design parameters to illustrate the wide range of turbine performance:

Inlet Conditions

Throttle	95,100 lb/hr at 615 psia, 800°F
Induction	22,100 lb/hr at 32 psia, 254°F
Exhaust Conditions	3.5 in. Hga

Extraction Flow and Turbine Output

Maximum	95,900 lb/hr	– 5710 kW at generator
Minimum	Zero	– 9980 kW at generator
Nominal	47,900 lb/hr	– 7910 kW at generator

Absorption Chiller Plant

The absorption chiller plant includes four absorption chillers with a total capacity of 5200 tons of refrigeration. The absorption chiller plant is designed to take individual absorption chillers out of service for part-load operation.

The absorption chillers are hermetically sealed vessels which contain several heat exchange sections where the steam, refrigerant, chilled water and cooling water heat transfer occurs. The steam is condensed in the chillers and collected in condensate receivers. The condensate is pumped through the domestic hot water heat exchanger to the deaerator.

PLANT PERFORMANCE

The cogeneration plant is designed for continuous base load operation 24 hours per day. The expected average plant performance is shown in Table 13-1. Also shown is the performance at full chiller load and zero chiller load. Efficiency, as calculated in accordance with PURPA regulations, is in excess of 45 percent.

Table 13-1
AVERAGE PLANT PERFORMANCE

	Average Performance	Chiller Full Load Performance	Chiller Zero Load
Net Electrical Power Output	27,900 kW	25,710	29,980
Chiller Load	2600 tons	5200 tons	0
Hot Water Heating	6.2×10^6 Btu/hr	11×10^6 Btu/hr	0
Fuel Consumption	267×10^6 Btu/hr	267×10^6 Btu/hr	267×10^6 Btu/hr
Fuel Utilization Efficiency	57.5%	76%	38%

SCHEDULING AND SITE CONSTRUCTION

Another true challenge of implementing the Dade cogeneration system is in the area of scheduling and construction. The building to house the cogeneration system had already been constructed while the permitting and contractual negotiations were in process. This Dade County Central Support Facility Building also houses office space, indoor parking, and additional building support systems.

Three 1500-ton centrifugal chillers are in service to supply the existing government buildings with chilled water prior to the cogeneration plant startup. These chillers will serve as backup as the cogeneration plant comes on line.

Figure 13-3 shows an artist's rendition of the cogeneration equipment arrangement inside the building. Fitting this sizeable equipment

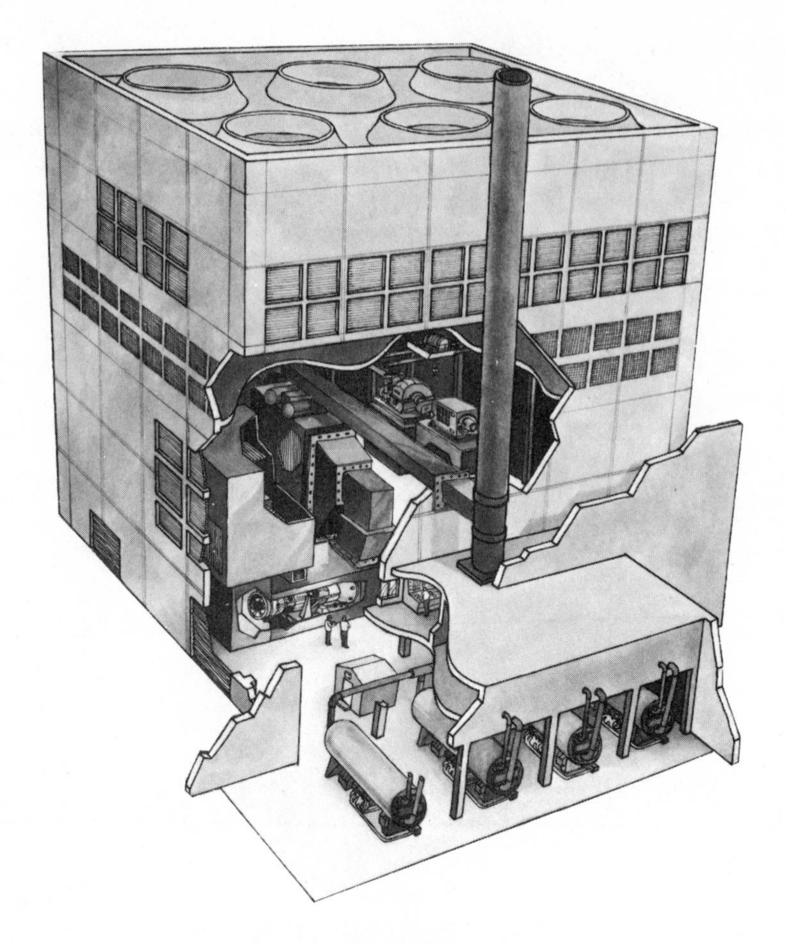

Figure 13-3
ARTIST'S RENDITION OF DADE COUNTY COGENERATION PLANT

in an existing structure takes careful planning and execution. An additional complication arises when it is considered that storage is unavailable at the site. Skid-mounted equipment was purchased to minimize field construction, and such equipment is off-loaded and put on prepared foundations in the building as expeditiously as possible. As an example, the large waste-heat boiler was delivered in two modules. The total erection time for this boiler was one week.

Large-bore piping, 2½ inches or greater, was purchased shop fabricated in spool pieces of field handleable sizes and is supplied with taps and/or nozzles as appropriate for receipt of instrumentation at the site. Such purchases alleviate the need for excessive field fabrication and construction scaffolding.

The Dade County schedule summary is shown in Figure 13-4. The total length of project is from 24 months to 30 months maximum. At times, engineering, procurement, and construction had been going on at the same time. Permitting was accomplished in phases to meet the schedules. Major equipment items had been purchased first. Foundations were then designed and, once completed, a general contractor was mobilized at the site. Thermo Electron engaged J.A. Jones Construction Company for the construction and installation of the cogeneration, absorption chiller and interbuilding distribution systems.

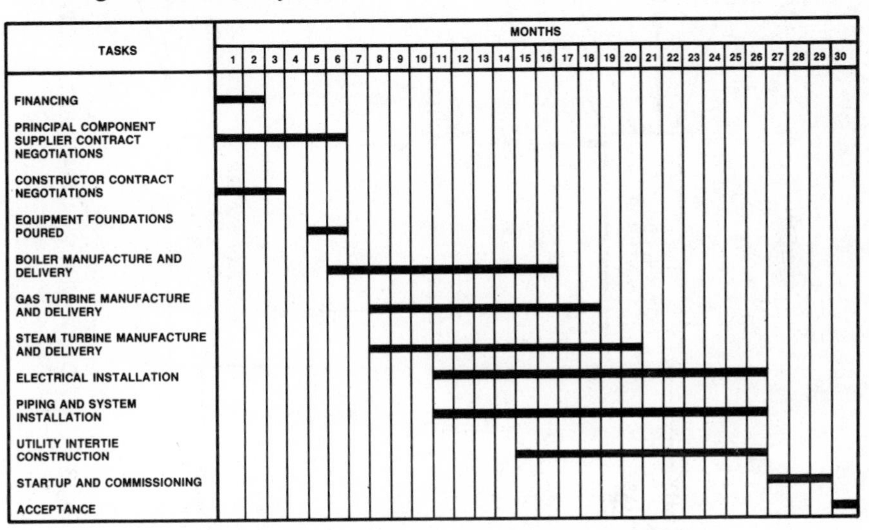

Figure 13-4 – DADE COUNTY SCHEDULE SUMMARY

FINANCIAL

This chapter would not be complete without mentioning the "financial engineering" involved in financing the $30 million Dade cogeneration plant. The following entities are participants in a third-party leveraged lease that could be the topic of several other presentations. (See Figure 13-5.)

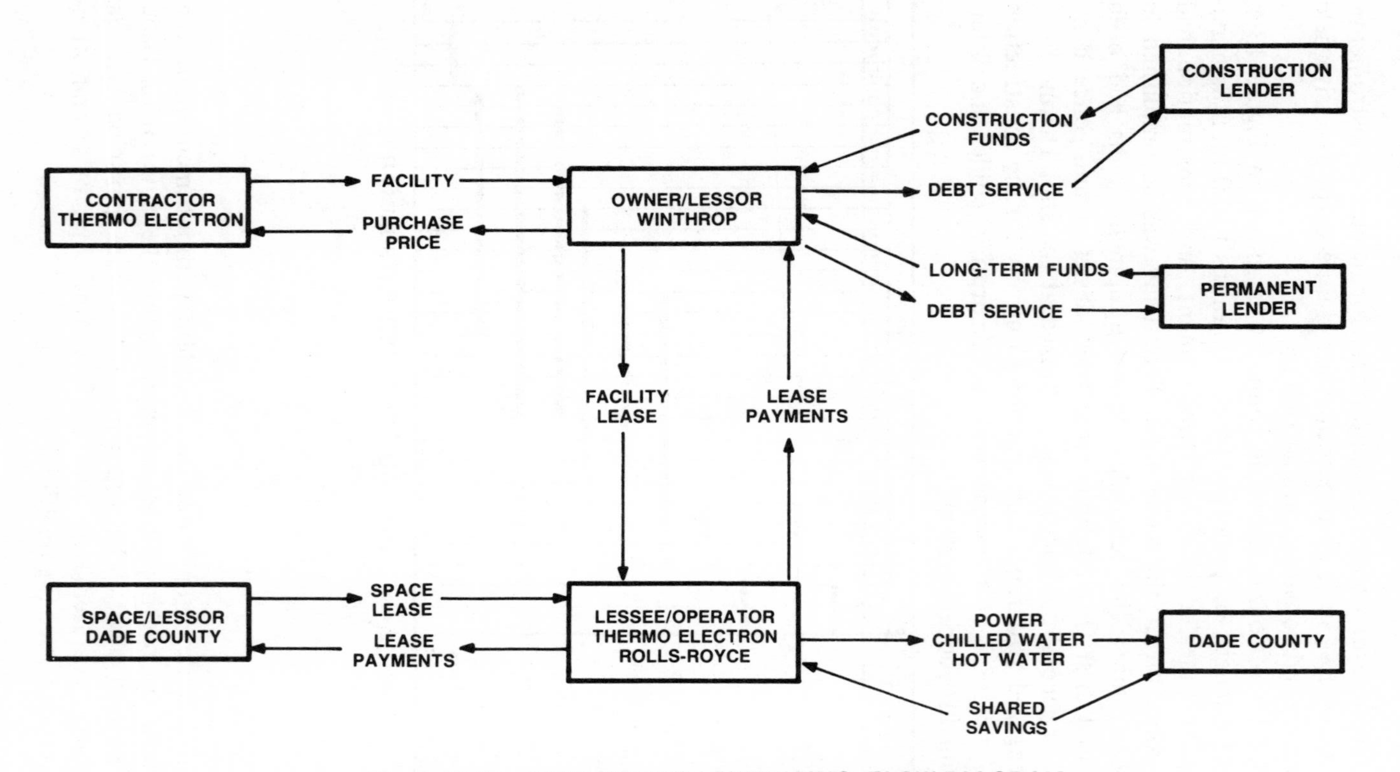

Figure 13-5 — DADE COUNTY FINANCING, FLOW DIAGRAM

- *Dade County*

  – The user of energy, in the form of electricity, chilled water and hot water from the joint venture and ultimate owner of the system after debt retirement.

- *South Florida Cogeneration Associates*

  The joint venture consisting of Thermo Electron's Florida subsidiary TEC Corporation and Rolls-Royce's Florida subsidiary RRD Corporation will operate the cogeneration system leased from Florida Energy Partners.

- *Thermo Electron*

  – Prime contractor and system designer for the cogeneration facility

  – Parent of TEC Corporation, a partner in South Florida Cogeneration Associates

- *Rolls-Royce*

  – Rolls-Royce Inc., parent of RRD Inc., a partner in South Florida Cogeneration Associates.

  – Rolls-Royce Ltd., the supplier of the gas turbine-generator set

- *Florida Energy Partners Limited Partnership*

  The partnership set up by Winthrop Financial Co.

  – Contracted with Thermo Electron Energy Systems for the construction of the cogeneration plant (the facility)

  – The lessor of the facility to the joint venture

SUMMARY

Cogeneration has a wide range of applications in municipal and institutional settings. When conceived and implemented with careful planning, projects can be accomplished in the most complex environments. Paramount to the successful completion of such projects is the choosing of collaborators who have the flexibility, depth, experience, and strengths to appreciate and respond to the technical, financial, legal, regulatory, political, and environmental issues involved.

CHAPTER 14

Technology Trends And Business Strategy For Cogeneration Systems

David Pasquinelli and Adil Talaysum

INTRODUCTION

The world has changed fundamentally in recent years and subsequently placed new burdens on the overall economic and political posture of every nation. The once stable supplies of premium energy resources are now disrupted by endogenous technical and exogenous political constraints. Foreign consortiums and cartels command strategic energy resources and threaten to compromise further the growth and stability of nations dependent on their supply. Sharply rising energy prices have not only changed our priorities, but stimulated the development of new energy conversion technologies.

The need to reduce the domestic consumption of foreign energy supplies, i.e., oil, has fostered the drive to utilize high efficiency energy conversion facilities that use domestic energy resources. The energy price difficulties of the early 1970s have changed the energy production equation and altered once established trends regarding economic growth and energy consumption. These changing parameters combined with changes in political emphasis and public perception have resulted in new industry and plant economies of scale and scope. The economics of electric power production have consequently been affected and the large central station concept challenged by new technologies.

On-site electrical generation is not a new concept, but has recently been resurrected as a viable supplement to large central stations.

In 1900, over 50 percent of the U.S. electric generating capacity was located at industrial sites. The electric utility industry was in its infancy and in many areas electricity was unavailable, unreliable and expensive. As the demand for electricity increased, economies of scale lowered the cost of central station power, technological advances improved its reliability and efficiency, and consequently, on-site generated power gradually decreased to 17% of total generating capacity in 1950, then 3% in 1980. New political, economic and regulatory trends have emerged that appear to be driving down the optimum power plant size and once again favoring smaller on-site generating plants. In the industrial sector, where approximately 40% of total electricity consumed is used and a substantial amount of thermal energy is also used, a new on-site power and heat producing technology known as Cogeneration has gained wide attention.

COGENERATION: TECHNOLOGY AND ECONOMICS

Cogeneration, in general terms, is defined as the sequential production of electrical or mechanical power and useful thermal energy from a single fuel source. Two cogeneration system arrangements are possible: topping and bottoming cycles. In topping cycles, the input energy is first used to produce power and the exhaust heat from the power-producing prime mover is used to generate steam or used directly in a heating process. Bottoming cycles are the reverse, utilizing waste heat from a heating process, such as an industrial furnace, to produce electricity. Commonly, the waste heat is used to produce steam in a heat recovery boiler and the steam is expanded through a steam turbine connected to a generator. An additional system arrangement is the combined cycle, in which the prime mover drives a generator, and steam produced in a heat recovery boiler is expanded through a steam turbine generator.

The advantage of cogeneration is that the overall cost of meeting an industrial facility's steam and electricity needs is less than if steam is generated in a boiler and electricity purchased from a utility. Typically, steam is produced in a separate boiler with an efficiency of 80-85%. Central station electrical power is produced in large Rankine cycle systems with an overall efficiency of 30-35%. Cogeneration systems simultaneously produce steam and electricity with an overall efficiency of 80-85%. Some congeneration units must use premium fuels such as natural gas and oil, whereas central stations and large boilers can use lower cost, non-premium fuels such as coal. Even with the fuel cost burden, the overall efficiency advantage of the cogeneration technology provides a strong economic incentive.

In order to assess the economic feasibility of a cogeneration opportunity, the following items must be considered:

- Amount of thermal and electrical energy consumed and usage profile
- Technologies utilized to provide the energy
- Fuel and electricity prices
- Special utility pricing features such as demand charge, supplemental power rates, and buy-back rates
- Projected fuel and electricity rate escalations
- Cogeneration system installed cost and performance
- Anticipated plant and/or process changes.

The following base case illustrates the economic viability of a typical cogeneration application. The following assumptions represent U.S. industry averages:

- natural gas cost: $4.50 per million Btu
- electricity cost: 6¢ per kWh
- steam cost: $6.50 per 1000 lb
- cogeneration equipment cost: $1000 per kW
- annual usage: 8400 hrs per year

One of the more promising cogeneration technologies is the gas turbine topping cycle. Several gas turbines are available in a wide range of sizes that provide good overall performance as cogeneration units. As an example, a gas turbine topping cogeneration unit producing 3 megawatts of electrical power also produces about 20,000 lbs per hr of 150 psig steam.

A typical industrial facility with utility needs in this size range would have the following energy cost break-down:

- steam cost: 20,000 lbs/hr ($6.50/1000 lbs) = $130/hr
- electricity cost: 3000 kW ($0.06/kWh) = $180/hr
- total energy cost: (130 + 180) (8400 hr) = $2,604,000/yr.

If a cogeneration system were installed to provide these energy streams instead of purchasing electricity from a utility and producing steam in a separate boiler, the energy cost and simple before tax pay-back would be as follows:

fuel cost:

(12,500 Btu/hr-kW) (3000 kW) ($4.50/MM Btu) = $170/hr

maintenance cost: $135,000/yr

total
cogeneration energy and maintenance cost: $1,563,000/yr

energy cost savings: $2,604,000 – $1,563,000
= $1,041,000/yr

$$\text{simple pay-back} = \frac{\text{Equipment cost}}{\text{Annual savings}}$$

$$= \frac{\$3{,}000{,}000}{\$1{,}041{,}000} = 2.88 \text{ yrs}$$

Generalized equations can be formulated to determine the economics of cogeneration. The full life-cycle costs associated with the investment can be reviewed, but for simplicity, capital costs can be modeled as annual rental rates (including the opportunity cost of capital, depreciation, taxes, etc.) and operating costs modeled as levelized values over the equipment life. In doing so, relatively simple equations can be derived that provide meaningful insight into fundamental cogeneration economic relationships.*

The total annual steam cost for a firm with conventional boiler technology is:

$$TC_b = K_bP + V_bPH$$

and for a firm with cogeneration technology is:

$$TC_c = K_cP + (V_c - V_e)\, PH$$

where: TC = total annual steam cost
K = annualized capital cost (rental rate)
V = fuel plus other operating costs per million Btu of processed steam produced

*The equations given below are derived from "Simple Economics of Industrial Cogeneration," *The Energy Journal,* January 1983.

P = uniform steam load in million Btu per hour
H = operating hours per year

note: subscript b = conventional boiler system
subscript c = cogeneration system
subscript e = electricity credit

The economic advantage of cogeneration is then the difference in cost between a conventional boiler and the cogeneration unit.

$$TC_b - TC_c = K_bP - K_cP + V_bPH - (V_c - V_e)\,PH$$
$$= (K_b - K_c)\,P + (V_b - V_c + V_e)\,PH$$

In general terms, the difference in total annual steam cost is a function of:

- the difference in capital cost, $(K_b - K_c)$
- the net energy cost balance for the boiler and cogeneration unit
- the operating hours per year.

To solve for the "break-even" running time for the cogeneration system, the above equation is solved for H when $TC_b = TC_c$:

$$H^* = \frac{K_c - K_b}{V_b - V_c + V_c}$$

As can be seen, the break-even running time increases as the cogeneration system becomes more expensive as compared to the boiler and decreases as the relative efficiency of the cogenerator increases. Packaging and integrating a cogeneration system to reduce installed equipment costs and increase overall system efficiency will reduce the break-even running time, thus making the unit more attractive.

In reviewing cogeneration system costs in general, capital costs tend to vary inversely relative to the quality of the fuel used. Fuels that command a premium price such as natural gas or distillate oil are utilized in cogeneration systems with the least expensive capital costs. Non-premium fuel such as coal, heavy residual oils and fuels derived from waste require expensive systems to be processed and, therefore, tend to be restricted to larger size systems.

The characterization of the plant utility load which encompasses the magnitude and variability of thermal and electric power requirements over time are important considerations in reviewing the economics of a cogeneration application. The ratio of thermal to electric requirements will influence the cogeneration technology selected, and the magnitude and variability will impact the size and control characteristics.

In evaluating a potential cogeneration application, a major consideration is to determine whether the system will be designed to:

- meet the total thermal and electricity needs at all times
- meet the thermal needs only at all times
- meet the electricity needs only at all times
- meet neither the thermal nor electricity needs at all times.

As the thermal and electricity demands increase in variability over time, the economic incentive of sizing the cogeneration system for the peak usage deteriorates. In the early beginnings of cogeneration, this strategy was attempted and contributed to the rapid decline of on-site power generation versus the central utility concept. Electric utilities face this seasonal load swing by sizing their non-premium fuel central stations for the steady base load seen throughout the year, and bringing premium fuel gas turbine units on and off line as needed throughout the annual cycle.

The cogeneration design strategy that has emerged favors matching the thermal load and having the electrical power production vary accordingly. In cases where the thermal load varies drastically, the cogeneration system is typically designed for the steady base load and auxiliary boilers used to carry the demand swings. To best utilize the variable electrical output of the cogeneration system, legislation was introduced* that allows the cogenerator to tie into the electrical utility grid and, therefore, obtain an "electricity credit" for electrical power supplied to the grid.

---

*This legislation, which was introduced to stimulate the use of cogeneration systems was implemented in 1978 as section 201 of the Public Utility Regulatory Policy Act (PURPA). This act stipulated that cogeneration facilities and small power production facilities which met certain standards and that were not utility owned were eligible for special incentive rates and treatment by the electrical utility.

EXISTING AND PROJECTED COGENERATION MARKET

There have been numerous studies performed that disaggregate the existing industrial cogeneration market and provide projections regarding future cogeneration opportunities. One of the most cited studies and comprehensive in its review of competitive cogeneration technologies was performed by the Office of Industrial Programs, U.S. Department of Energy.*

Table 14-1 illustrates a profile of existing industrial cogeneration by industry. Currently, in the industrial market there is approximately 12,000 megawatts of cogenerated electrical capacity produced in 471 plants. Based on generated capacity, the Paper industry is the largest sector, followed by the Primary Metals, Chemicals, Petroleum, Other Manufacturing and finally, Food. The Paper industry also has the largest number of plants, but the number of plants does not follow the same order for the list of industrial sectors arranged in order of decreasing generation capacity. The average generation capacity per plant varies for each industrial sector as shown in Table 14-1:

Table 14-1
PROFILE OF EXISTING INDUSTRIAL COGENERATION BY INDUSTRY

Industry	Generation Capacity (MW)	Percent	Number of Plants	Average Plant Size (MW)
Food	409	3.4	48	8.5
Paper	3764	31.3	189	19.9
Chemicals	2429	20.2	86	28.2
Petroleum	1115	9.2	32	34.8
Primary Metals	3643	30.3	41	88.9
Other	671	5.6	75	8.9
Total	12031	100.0	471	25.5

*U.S. Department of Energy, *Industrial Cogeneration Potential: Targeting of Opportunities at the Plant Site,* DOE/CS/4062, May 1983, and *1985 Update of Industrial Cogeneration Potential,* 1985.

It is interesting to note that the average capacity per plant varies from a high of 88.9 megawatts per plant to a low of 8.5 megawatts per plant. Also, the industrial sectors that have experienced the most rapid penetration of cogeneration capacity are those with the highest average capacities per plant. As would be expected, the larger cogeneration applications utilizing relatively inexpensive nonpremium fuels were the first to be exploited. As an example, the Paper industry is a leader in cogeneration due to the large steam and power requirements and large amounts of burnable process waste typically available in an integrated pulp and paper mill. The synergistic match between utility requirements and fuel supply, combined with the magnitude and consistent supply and demand of each, have created strong economic incentives and subsequently rapid technology penetration in the Paper, Chemicals, Petroleum and Primary Metals sectors. At the other end of the scale, 26% of the plants are located in the Food and Other Manufacturing Sectors, which make up only 9% of the generated capacity. Apparently, these industries are structured to accommodate a large number of small plants or plants utilizing premium fuels, which also tend to favor the smaller sized systems.

The projected industrial cogeneration market based on the DOE study is shown in Table 14-2, which presents the forecast capacity and number of plants in the top 7 industrial sectors. The study fore-

Table 14-2

PROJECTED INDUSTRIAL COGENERATION MARKET

SIC	Industry	Capacity (MW)	%	Number of Plants	%	Average Size (MW)
20	Food	7005	18	734	20	9.5
22	Textile	1882	5	499	14	3.8
26	Paper	7962	20	437	12	18.2
28	Chemicals	10,316	26	547	16	18.9
29	Petroleum	5836	15	236	6	24.7
33	Primary Metals	2397	6	434	12	5.5
	Other Manufacturing	3950	10	730	20	5.4
	Total ...	39,348	100	3,644	100	10.8

casts a total cogeneration capacity of 39,348 megawatts in 3,644 plants. The data is fairly consistent and it is interesting to note that the aggregate industrial plant average size has decreased.

Table 14-3 illustrates the projected industrial cogeneration market by system size. The 20 to 50 megawatts size range is the largest group based on capacity, and the 2 to 5 megawatt size range is the largest group based on number of plants. Table 14-4 illustrates the 13 technologies considered vs. forecasted number of plants and generating capacity. This indicates that coal-fired steam turbine cogeneration systems make up the largest group based on generating capacity, and natural gas-fired gas turbines are the most popular technology when considering number of plants.

Table 14-3
PROJECTED INDUSTRIAL COGENERATION MARKET

System Size (MW)	Total MW Production	MW %	Number of Plants	Plants %
<1	295	0.7	416	11.4
1-2	867	2.2	595	16.3
2-5	3643	9.3	1088	29.9
5-10	4201	10.7	605	16.6
10-20	6634	16.8	462	12.7
20-50	10247	26.0	334	9.2
50-100	7621	19.4	108	3.0
100-200	3487	8.9	27	0.7
>200	2355	6.0	9	0.2
	39348	100.0	3644	100.0

COGENERATION MARKET BARRIERS

The cogeneration market presented thus far is an optimistic projection based on relatively low economic thresholds. As the perceived uncertainties and risks increase, higher economic returns are required that subsequently limit the pool of viable opportunities. Understanding the technological, institutional and economic barriers to cogeneration is instrumental in understanding the potential market.

Table 14-4
POTENTIAL INDUSTRIAL COGENERATION MARKET

Prime Mover/ Fuel	Number of Potential Cogeneration Plants	%	Potential Power Generation (MW)	%
Diesel/natural gas	3	0.1	4	0
Diesel/residual	11	0.3	14	0
Diesel/distillate	0	0	0	0
Steam turbine/natural gas	24	0.7	1162	3.0
Steam turbine/residual	80	2.2	2679	6.8
Steam turbine/distillate	2	0.1	61	0.2
Steam turbine/coal	697	19.1	19633	49.9
Gas turbine/natural gas	1797	49.3	7496	19.1
Gas turbine/residual	310	8.5	5842	14.8
Gas turbine/distillate	715	19.6	2360	6.0
Combined cycle/natural gas	0	0	0	0
Combined cycle/residual	1	0	37	0.1
Combined cycle/distillate	4	0.1	61	0.2
	3644	100.0	39348	100.1

The following list of identified cogeneration market barriers is believed to be the most comprehensive list available from any reference. The identified barriers were determined as a result of a thorough literature search and direct discussions with end users, equipment manufacturers, and government officials.

- *Regulatory Uncertainty* – The key regulations that qualify and regulate cogenerators are relatively new and are still being challenged by the utility industry. PURPA, the most central regulation concerning cogeneration, has weathered several court challenges but still has several provisions subject to interpretation. Changes in government policy concerning other investment incentives such as "energy tax credits" also greatly affect the market.

- *Stricter Air Quality Standards* – Pollution regulations recently introduced in California could substantially add to the initial capital investment required for a cogeneration unit. The increased capital cost and, in some cases, reduced system performance, will discourage the use of certain cogeneration technologies.

- *Changing Utility Buy-Back Rates* – Electric utility buy-back rates are negotiated for each cogeneration site and are determined by the utility's avoided costs at that point in time. As the structure of the utility industry changes and as more cogeneration capacity is brought on line, these buy-back rates may decrease and therefore discourage further cogeneration.

- *A Reluctance to Enter the "Power Business"* – Most manufacturing concerns have become accustomed to simply purchasing power and generating steam or process heat on-site. The prospect of adding a cogeneration system represents a large capital investment in what is usually an unfamiliar technology.

- *Electric Utility Resistance* – Many electric utilities have adequate or excessive capacity reserve margins, and therefore, have little incentive to purchase electrical power from a cogenerator. Other utilities disagree with the distributed on-site power concept, having grown up during a period when large central stations dominated the utility structure.

- *Lack of Cogeneration Expertise* – Most industrial concerns do not have in-house personnel capable of evaluating, operating or maintaining cogeneration systems. In general, cogeneration technology expertise has been confined to specific industries that have traditionally utilized equipment of this type.

- *Perceived Unreliability of Gas Turbines* – Gas turbines, which are typically perceived as high speed and sensitive equipment, are not generally utilized in many industrial prime mover applications. Most gas turbines are used in power producing applications as peaking generator sets, and do require frequent maintenance due to the severe start-up and shut-down operating cycle.

- *Limited Capital Resources* – Many industrial concerns do not have sufficient internally generated funds, or are unwilling or unable to obtain external funds.

- *Higher Priority of Production Oriented Projects* – Cogeneration projects are oriented towards energy savings, and therefore, take a lower priority than production oriented projects. Higher returns on investment are typically required with energy saving projects when competing for limited capital budgets with production projects.

- *Stability of Energy Prices* – Recent fluctuations in the energy market have created a sense of uncertainty with regard to long-term energy price projections. Small variations in the input and output utility costs of a cogeneration unit could easily destroy the economics of the project.

- *Possible Plant Closings* – Companies will limit the scope of capital investment targeted at plants that may be severely curtailed or closed as a result of changing business or economic climates. Cogeneration projects typically require years of steady operation to be economically viable, and therefore, are of limited interest in this situation.

- *Premium Fuel Availability* – Only recently have natural gas and oil been available in adequate supplies and reasonable prices to the industrial sector throughout the entire year. The long-term availability of reasonably priced premium fuel is crucial to the successful implementation of several cogeneration technologies.

- *Limited Floor Area* – Most cogeneration systems are retrofitted into existing manufacturing facilities with limited available floor space. Adequate plant volume for the proper operation and maintenance of the cogeneration unit in an area that is accessible to the plant utility tie-ins are important considerations.

- *High Capital Cost for Cogeneration Systems* – The available prime mover technologies limit the potential sizes of cogeneration systems to relatively large and expensive units. The high initial capital cost of cogeneration systems is a barrier in industries with limited financial strength.

- *Proper Size and Thermal to Electric Output Match* – Accommodating the application's energy stream requirements with that available from the various cogeneration technologies is instrumental to a successful project. Most cogeneration technologies have relatively fixed thermal to electric outputs and most applications have daily and/or seasonal variations in utility requirements.
- *Adequate Water Supply* – Many cogeneration technologies utilize heat recovery boilers that require a steady supply of high quality feedwater.
- *New Production Technologies that Reduce Energy Usage* – New manufacturing technologies are emerging that could drastically reduce the utility requirements of current industrial processes. This potential change in energy usage or change in the composition of energy required could alter the viability of certain cogeneration technologies.

Understanding the barriers that a new technology faces in the market is the first step in developing an effective project strategy. The identified market barriers encompass a wide range of business, government policy, institutional, regulatory, financial, environmental, economic, and technology problems. In reviewing the barriers and market data, there appears to be a unique need for small, packaged gas turbine cogeneration systems.

COGENERATION BUSINESS STRATEGIES

The total projected market for cogeneration is nearly 40,000 megawatts of capacity. Most firms currently active in the cogeneration market focus their efforts on a particular cogeneration technology or specific industrial applications. The major equipment manufacturers and engineering firms were first to enter the cogeneration market and have concentrated on the large cogeneration applications over 20 megawatts capacity. These projects were addressed in a typical A&E fashion, based on their experience in working with large central station plants, and involved a great deal of custom engineering, field fabrication and erection of the cogeneration plant.

Large-scale cogeneration projects could absorb the overhead involved with a construction engineering firm, but this design and construction approach is uneconomical for cogeneration projects under 10 megawatts capacity. As previously illustrated in the projected industrial cogeneration market data, four of the seven industrial sectors (food, textile, primary metals and other manufacturers) have average plant capacities less than ten megawatts. These four industrial sectors encompass 39% of the projected cogeneration capacity and 66% of the number of plants. This significant share of the projected cogeneration market for smaller capacity cogeneration units is basically not economically viable if the typical large engineering and construction firm is used to supply these plants.

To further illustrate the importance of these four industrial sectors, the potential for cogeneration market growth was determined by calculating the ratio of projected to existing industrial cogeneration capacity and number of plants per each industry as shown in Table 14-5.

Table 14-5
RATIO OF PROJECTED TO EXISTING INDUSTRIAL COGENERATION MARKET

Industry	Capacity	Number of Plants
Food	17.1	15.3
Paper	2.1	2.3
Chemicals	4.3	6.4
Petroleum	5.2	7.4
Primary Metals	0.7	10.6
Other Manufacturing	5.9	9.7
Industry Average . . .	3.3	7.7

The Food, Primary Metals, and Other Manufacturing sectors exhibit the highest potential for growth when considering number of plants of any industrial sector and, except for the Primary Metals sector, exhibit the highest capacity growth as well. Table 14-6 illustrates the purchased and on-site generated electricity profile for each of the industrial sectors. As of 1980, only 8.2% of all electricity used by these industrial sectors was generated on-site and the Food, Primary

Table 14-6
1980 INDUSTRIAL MANUFACTURING ELECTRICITY CONSUMPTION

SIC	Industry	Purchased Electricity (Billion kWh)	Generated Electricity (Billion kWh)	Percentage Generated On-Site
20	Food	41.1	2.2	5.1
26	Paper	49.7	26.3	34.6
28	Chemicals	133.2	11.8	8.1
29	Petroleum Refining	32.2	5.5	14.6
32	Stone, Clay & Glass	30.5	0.3	1.1
33	Primary Metals	164.2	11.2	6.4
	Other Manufacturing	202.2	1.5	0.7
	Total	658.1	58.8	8.2

Source: 1980 Annual Survey of Manufacturers

Metals and Other Manufacturing sectors generated on-site only 5.1, 6.4, and 0.7 percent, respectively, of the total electricity consumed. The Other Manufacturing sector purchased the largest amount of electricity, 202.2 billion kilowatt hours while generating on-site only 1.5 billion kilowatt hours, and therefore, represents a substantial market opportunity for cogeneration. The projected market and industry generation profile data strongly indicates a substantial market in industrial sectors characterized by smaller capacity cogeneration systems.

In reviewing and selecting specific industrial sectors for cogeneration, the overall growth projections for the sectors are important. Table 14-7 illustrates the industrial growth projections for each of the industrial sectors discussed, where it can be seen that the Chemicals and Other Manufacturing sectors have the highest growth potential at approximately 4% per year followed by the Paper, Petroleum, Primary Metals, Food and Textile industrial sectors. Of the industrial sectors favoring smaller sized cogeneration units, the Other Manufacturing sector has the highest growth potential and is further broken down by subsectors in Table 14-7. Of the 6 subsectors illustrated, the Electric and Electronic Equipment, Instruments and Machinery subsectors have overall growth potential over 4% per year. These strong overall growth projections and cogeneration potential clearly illustrate initial targets of opportunity.

Table 14-7
INDUSTRIAL GROWTH PROJECTIONS

Industry	Growth (% Per Year)
Food	1.7
Textile	1.4
Paper	3.4
Chemicals	4.2
Petroleum	2.5
Primary Metals	1.8
Other Manufacturing	4.1
Fabricated Metal Products	2.5
Machinery	4.0
Electric and Electronic Equipment	7.1
Transportation Equipment	2.3
Instruments and Related Products	6.0
Misc. Manufacturing	2.5

Having refined the market and identified an opportunity for smaller sized cogeneration units in specific industrial sectors, the specific technology to be focused in the target market must be determined. Previously, it was indicated that the smaller capacity systems favor technologies utilizing premium fuels. Also, the cogeneration unit should be sized in the 3 to 5 megawatt size range to properly address the identified applications. In reviewing Table 14-4, gas turbine cogeneration units using natural gas, residual or distillate oil make up a large percentage of the total potential plants, and if the average capacity per plant is determined, fulfill the target size desired as shown in Table 14-8. As illustrated, natural gas and distillate oil gas turbine cogeneration systems exhibit an average capacity per plant of 4.2 and 3.3 respectively. Residual oil, which is a heavier more difficult fuel to process, has an expected higher capacity per plant of 18.8.

Gas turbine cogeneration systems which drive a generator from the gas turbine and produce steam from the turbine exhaust gases are compact, exhibit overall system efficiencies greater than 80%, and have relatively high electric to thermal energy output. The initial market targets are primarily in the manufacturing sector and the

Table 14-8

Technology	**Number of Plants**	**Capacity (MW)**	**Average Capacity Per Plant**
Gas Turbine			
•Natural Gas	1797	7496	4.2
•Residual Oil	310	5842	18.8
•Distillate Oil	710	2360	3.3
	2817	15,698	5.6

gas turbine cogeneration system closely matches the energy requirements and overcomes many of the barriers identified in this market.

Figure 14-8 outlines the principal sub-systems that make up a gas turbine cogeneration system. As can be seen, the system is composed of two primary subsystems, the gas turbine generator system and the heat recovery boiler system. To enhance steam production, duct burner systems are commonly incorporated and can provide twice the steam produced in the unfired system.

Critical to the overall design of the cogeneration system is the selection of the gas turbine. Seven domestic gas turbines are available in the 3 to 5 megawatt size range as follows:

Gas Turbine Engine	**Nominal Rating (kW)**
Solar Centaur	2800
Allison 501-KB	3066
Rustin TB5000	3460
GE LM500	3600
Dresser 990	4327
Allison 570-KA	4500
Allison 571-KB	5100

Tables 14-9 and 14-10 provide additional data on each of the gas turbines and illustrate the calculation of a simple payback to compare the turbines based on the electric and steam income and fuel costs only. The key parameters in evaluating a gas turbine for cogeneration usage are the heat rate or efficiency, turbine exhaust gas temperature, and turbine dry weight. Basically, the lower the heat rate or higher the efficiency, the higher the exhaust gas temperature, and the lower the turbine weight, the more desirable a gas turbine is for cogeneration applications.

Table 14-11 compares the gas turbines with respect to electric and steam income, fuel costs, maintenance costs and overall economic parameters such as simple payback, return on investment, net present value and profitability index.

The overall technology strategy is to package the units so that the system can be totally factory assembled and tested therefore minimizing expensive field erection. This design approach will improve efficiency, enhance reliability, reduce the overall unit cost, and minimize the floor area required in the plant. A system utilizing this design philosophy could be provided at less than $800 per kilowatt installed equipment costs and be delivered to the plant site in less than one year.

CONCLUDING REMARKS

The overall posture of the energy market and disenchantment with large central station power systems has created a substantial opportunity for cogeneration systems. Unique opportunities appear to exist for small capacity gas turbine topping cogeneration systems in several industrial sectors. The technological and economic benefits are currently attractive and will become more so as electricity prices continue to rise and natural gas and oil prices stabilize. A growth spurt in the economy will begin to strain the existing electric utility capacity, which will enhance their cooperation with cogenerators. Cogeneration can service an important national need and improve the overall efficiency of the energy sector.

Table 14-9
COMPARISON OF GAS TURBINE SYSTEMS

Gas Turbine Engine (Year Introduced)	Nominal Rating (kW)	Heat Rate (Btu/kW-hr)	Air Flow (lb/sec)	Turbine Exhaust (°F)	Dry Weight (lb)	Steam Production (lb/hr)	Nat. Gas Usage (MM Btu/hr)	Net Income ($1000/yr)	Simple Payback (yrs)
Solar Centaur (1968)	2800	13,930	38.0	813	40,000	17,290	39.00	881.2	3.18
Allison 501-KB (1962)	3066	12,844	33.5	950	1270	19,750	39.38	1135.7	2.70
Ruston TB5000 (1977)	3460	14,150	46.0	944	30,000	26,850	48.96	1359.1	2.55
GE LM500 (1980)	3600	11,537	34.0	940	2260	19,710	41.53	1320.5	2.73
Dresser 990 (1978)	4327	11,305	43.0	880	7500	22,390	48.92	1554.0	2.78
Allison 570-KA (1979)	4500	12,150	42.8	1150	1350	33,640	54.68	2037.8	2.21
Allison 571-KB (1982)	5100	11,400	44.2	1070	1490	31,270	58.14	2080.7	2.45

Table 14-10
COMPARISON OF GAS TURBINE SYSTEMS

Gas Turbine Engine	Nominal Rating (kW)	Steam Production (lb/hr)	Electric Income ($/hr)	Steam Income ($/hr)	Fuel Cost ($/hr)	Net Income ($/hr)	Net Income ($1000/yr)	Simple Payback (yrs)
Solar Centaur	2800	17,290	168.0	112.4	175.5	104.9	881.2	3.18
Allison 501-KB	3066	19,750	184.0	128.4	177.2	135.2	1135.7	2.70
Ruston TB5000	3460	26,850	207.6	174.5	220.3	161.8	1359.1	2.55
GE LM500	3600	19,710	216.0	128.1	186.9	157.2	1320.5	2.73
Dresser 990	4327	22,390	259.6	145.5	220.1	185.0	1554.0	2.78
Allison 570-KA	4500	33,640	270.0	218.7	246.1	242.6	2037.8	2.21
Allison 571-KB	5100	31,270	306.0	203.3	261.6	247.7	2080.7	2.45

Table 14-11

Gas Turbine Engine	Nominal Rating (kW)	Electric Income ($1000/yr)	Steam Income ($1000/yr)	Fuel Cost ($1000/yr)	Maintenance Cost ($1000/yr)	Depreciation Cost ($1000/yr)	Net Income ($1000/yr)	Simple Pay-Back (Yrs)	ROI (%)	NPV ($1000)	PI ($1000)
Solar Centaur	2800	1411.2	944.2	1474.2	140.0	280.0	461.2	6.1	16.5	39.9	1.012
Allison 501-FB	3066	1545.6	1078.6	1488.5	153.3	306.6	675.8	4.5	22.0	1086.5	1.354
Ruston TB5000	3460	1743.8	1465.8	1850.5	173.0	346.0	840.1	4.1	24.3	1702.1	1.492
GE LM500	3600	1814.4	1076.0	1569.9	180.0	360.0	780.5	4.6	21.7	1195.9	1.332
Dresser 990	4327	2180.6	1222.2	1848.8	216.4	432.7	904.9	4.8	20.9	1233.4	1.285

Notes:

1. Maintenance Cost = 5% of installed equipment cost
2. Depreciation Cost = 10% of installed equipment cost
3. Cost of Capital = 10%
4. Project Life = 10 years

CHAPTER 15

Combined Cycle Cogeneration At Nalco Chemical

Cabot B. Thunem, P.E.,
Kenneth W. Jacobs, P.E., and William Hanzel

INTRODUCTION

The Nalco Chemical Company is a Fortune 500 company with annual sales of more than 660 million dollars. Nalco is an international producer of specialty chemicals and provides products and services worldwide for water and waste treatment, pollution control, petroleum production and refining, papermaking, mining, steelmaking, metal working, and other industrial processes.

In 1977, Nalco purchased acreage in Naperville, Illinois, to build a 250,000 sq ft Technical Center and as a future site for an office complex. The initial facility, completed during the fourth quarter of 1979, consisted of an administration building, three laboratory buildings, a powerhouse, and parking areas. In November, 1983, construction of a 400,000 sq ft office complex was started. The new construction includes three connected office buildings, an addition to the powerhouse, and additional parking. These new facilities are scheduled for completion in March, 1986. The new complex will house the headquarter offices and training center currently located in Oak Brook, Illinois. After March 1986, Nalco will have about 1,150 people housed at the Naperville site.

Construction on the site is progressing. Currently, all building exteriors have been completed and interior work is underway. The project is on schedule.

During the Technical Center's existence, utility costs have increased steadily, and periodic problems with power outages and brownouts have occurred. Currently, all energy used within the

Technical Center is purchased: natural gas from Northern Illinois Gas Co. and electric power from Naperville Electric. Since 1980, combined utility rates have been increasing by more than 12% per year. In 1986, Nalco expects the annual cost of utilities for the Technical Center to be 2.0 million dollars. With occupancy of the office buildings in 1986, these costs could be 4.0 million dollars. Assuming a 6% per year escalation rate, an extrapolation of these costs indicates that by 1995 the annual utilities bill could exceed 6.0 million dollars for gas and electricity.

Like many chemical companies, Nalco routinely investigates backward integration relating to the possible production of raw materials for their own use. Construction of Corporate offices at the Naperville site spurred the investigation of another type of backward integration: the on-site production of electricity. To maximize economics of on-site production, cogeneration, rather than simple cycle power production, was investigated. Preliminary economic analysis of this system configuration indicated sufficiently attractive returns; consequently, a formal feasibility study was initiated, and included the following steps:

1. Feasibility Study
2. Design Outline
3. Detail Design
4. Construction.

COGENERATION APPROACH

Three modes of cogeneration are typically available: steam cycle, gas turbine, and reciprocating engine. Preliminary analysis indicated that neither steam cycle cogeneration nor reciprocating engine cogeneration would provide the optimum solution to on-site cogeneration of electricity and steam. Combined cycle cogeneration with gas turbines would probably produce superior returns to simple cycle cogeneration with gas turbines.

The first step in the feasibility study is establishment of the parameters necessary for sizing the system, and evaluation of project economics. The parameters identified include:

- fuel cost
- electrical cost

- required capital
- electric to steam use ratio
- cost of funds
- PURPA regulations
- TEFRA regulations
- usage patterns.

Before system sizing or system configurations could be established, a daily steam and electric-use pattern was required for both the existing system and the extrapolated consumptions for the expanded facility. Feasibility for cogeneration then proceeded with the following three steps:

1. Determine the capital cost and performance of each separate cogeneration design alternative.
2. Perform economic and sensitivity analyses of significant cost parameters.
3. Determine environmental and regulatory compatibility of the proposed systems with state and federal regulations.

SYSTEM CONFIGURATION SELECTION

Although a number of technologies may be applied to the cogeneration concept, natural gas combustion turbines with waste heat recovery were selected. Combustion turbines were selected because of their high overall efficiency, their ability to match the electrical and steam demands of the Nalco facility, and their simplicity of operation.

A number of factors require evaluation in the overall economic analysis for cogeneration in any facility. Each of the following factors must be included in considering sizing of the system, selection of the particular turbine, and design of the waste heat steam generator:

Steam Load

Current steam use pattern determined from operating records indicated that high steam rates were needed in winter for heating and very little steam was required in summer.

Current summer cooling is provided by electric chillers. Nalco elected to install an absorption chiller to handle the cooling load for the new offices. This provides the summer steam load needed to support the proposed turbine system. The heat loads for the new complex were estimated and added to current requirements. The plot of this data (Figure 15-1) indicated the project could support a system having output of 15,000 to 20,000 lb/hr during the summer and about twice that during the winter.

Electric Load

The same procedure was used to determine the daily electric use pattern. A plot of this data showed that the expanded facility could support a system having an electric output of 3.0 to 4.5 MW (Figure 15-2).

Economic and Financial Analysis

Following completion of the process development for each configuration, options were evaluated through economic modeling.

Financial analysis projects capital costs, gas consumption, operating costs, and electric and steam output into an economic model to simulate the financial performance of the system over a 20-year period. Modeling is performed for each configuration. The model provides a projection of each year's anticipated benefits and costs, which are then converted into summary statistics (internal rate of return, net benefit, and savings investment ratio) for purposes of comparing one alternative with another.

Regulatory Incentives

In addition to the benefits of high thermal efficiency, several tax incentives are provided to cogenerators. The Tax Equity and Fiscal Responsibility Act of 1982 (TEFRA) provides a 5-year accelerated depreciation as well as a 10% investment tax credit. The Public Utility Regulatory Policies Act (PURPA) requires utilities to interconnect with cogenerators and to provide maintenance power, backup power, and supplementary power without penalty by the utility. In addition, cogenerators may sell excess power at the negotiated avoided cost to the utility. These factors were incorporated into the financial model.

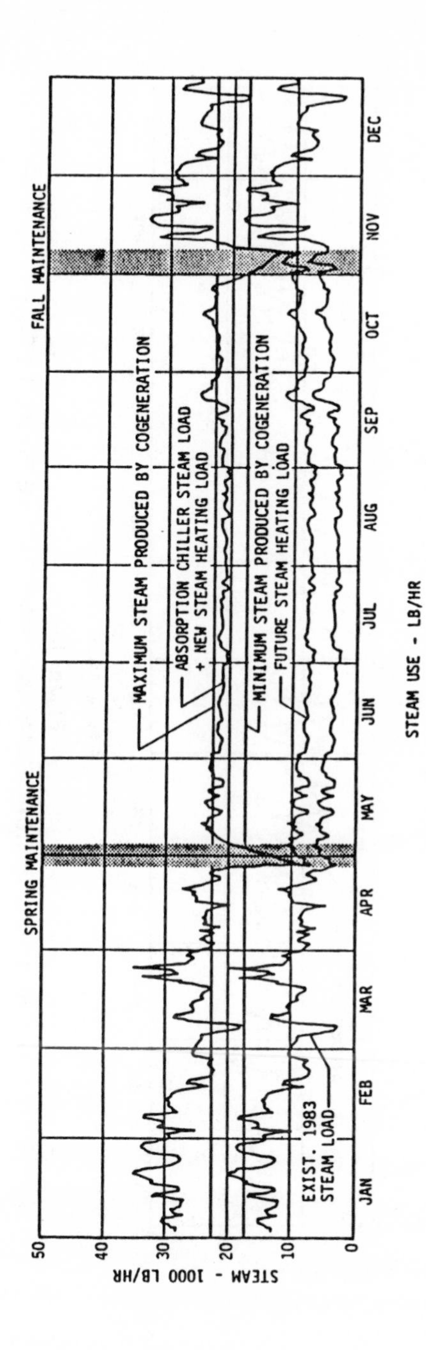

Figure 15-1
NALCO CHEMICAL COMPANY, NAPERVILLE FACILITIES
STEAM USE PROFILE

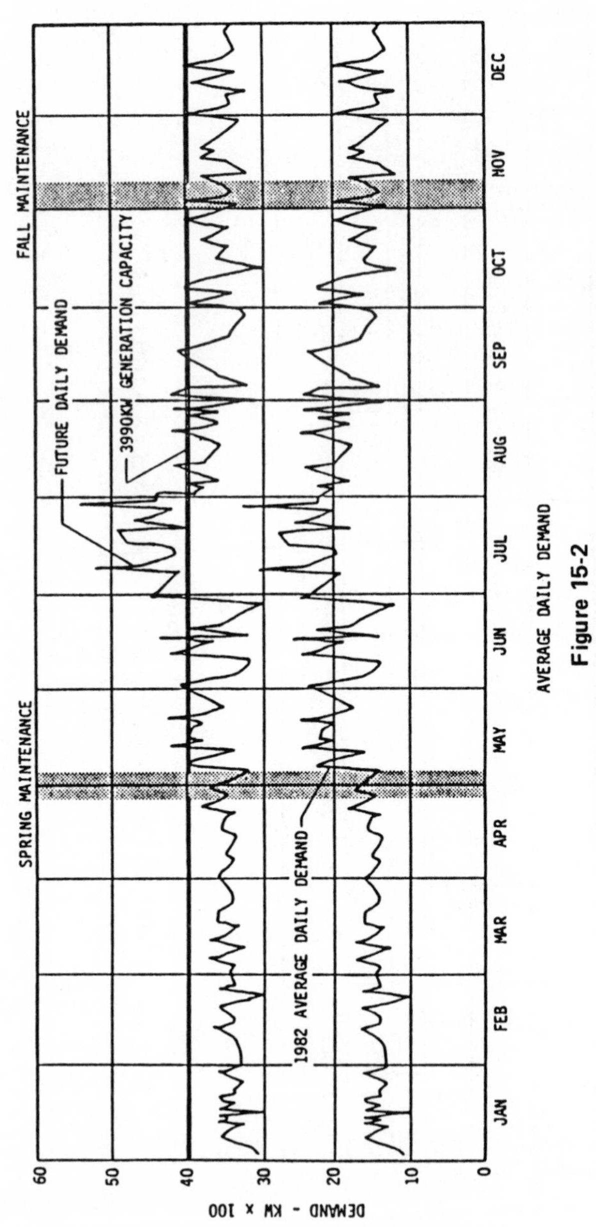

Figure 15-2
NALCO CHEMICAL COMPANY, NAPERVILLE FACILITIES
AVERAGE DAILY POWER DEMAND

Environmental Regulations

Several guidelines and regulations have been promulgated on the federal and state levels for mitigation of environmental impacts resulting from combustion. The primary environmental concerns include air emissions, water uptake and discharge, and solid waste disposal. The proposed gas turbine cogeneration installations for Nalco should have no impact on water intake or discharge, nor on solid waste disposal. The only air emission resulting from the proposed cogeneration facility is exhaust from the gas turbine.

Exhaust emission from the facility is subject to state and federal regulations. New stationary gas turbines greater than 10.7 gigajoules per hour (10 million Btu per hour) are regulated by Federal New Source Performance Standards (NSPS). These standards restrict the emissions of nitrogen oxides and sulfur dioxide. The nitrogen oxide standard is based on the formula: 0.015 times (14.4 ÷ heat rate at rated peak load in kilojoules per watthour) + an allowance for fuel-bound nitrogen. This emission rate is a percentage by volume at 15% oxygen on a dry basis.

The sulfur dioxide standard limits emissions to 0.015% sulfur dioxide by volume or a sulfur content of the fuel less than 0.8% by weight.

In addition to these emission limits, the NSPS regulations require owners or operators of the new source to monitor the sulfur and nitrogen content of the fuel fired into the turbine. If water or steam injection is used, owners or operators must continuously monitor the fuel consumption and the ratio of water to fuel being fired. The frequency of monitoring is specified in the Code of Federal Regulations, Part 60, Subpart GG.

Since the NSPS for gas turbines have been in existence since 1979, most manufacturers have designed their gas turbines accordingly. However, compliance with these regulations had to be demonstrated to the Illinois Environmental Protection Agency, which administers the NSPS in Illinois.

REGULATORY CONSIDERATIONS

The Tax Equity and Fiscal Responsibility Act of 1982 (TEFRA) – Signed into law on September 3, 1982, this act is a key element in financing a generation facility. This legislation, along with other federal tax laws, provides guidance regarding tax benefits for a project of this type.

All generation equipment for the project is depreciable over a 5-year period. Buildings and any structural components do not qualify for the 5-year depreciation period. The regular investment tax credit rate is 10% (Tax Code Section 46). This tax credit applies to equipment, but not to the building or structural components.

Public Utilities Regulatory Policies Act of 1978 (PURPA) – Removed many of the regulatory constraints from cogenerators and defined responsibilities of utilities in purchasing power from firms that are not primarily in the generation business. PURPA extends several important incentives to qualifying cogenerators and small power producers. The incentives include:

- Exemption from electrical utility regulation.
- Incentive rates for sales on purchase of electricity.
- Special provisions on interconnection.

PURPA requires electric utilities to purchase power from cogenerators and small power producers at a rate that does not exceed the incremental alternative electric energy cost to the electric utility. This is termed the utility's "avoided cost" and is measured by the savings to the utility in not generating the power itself or purchasing from the grid.

Several court challenges were raised over the avoided cost principle as administered by the Federal Energy Regulatory Commission. After an appeals court struck down the avoided cost concept in 1982, the U.S. Supreme Court overturned the decision in mid 1983. The court ruled that the avoided cost rule was in full force, provided that the two parties had not entered into an agreement as to price and rate structure. In actual practice for long-term purchase agreements, the price per kWh may be only 80%-90% of the full avoided cost.

The Fuel Use Act (FUA) – This act prohibits the use of oil or natural gas as a primary energy source in 1) any new electric power plant, and 2) any new major fuel burning installation that consists of a boiler, (nonboilers, including gas turbines are not covered) unless an exemption is granted.

Utility Interconnects – Provision has been made for connection of the generator to the utility grid, including necessary transformations for connection to the plant power bus and the electric utility substation. The gas turbine-generator will operate at the nominal 12,500-volt substation delivery voltage. Interconnection is accomplished under PURPA guidelines.

The generator will be equipped with metering, instrumentation, and controls to permit synchronization and parallel operation with the electric utility. The generator is capable of supplying power to the plant system as well as back into the utility system.

The generator required a complement of relays for protection of the unit in the event of electrical or mechanical malfunctions. Provisions have been made for supply connections from the distribution system to the cogeneration auxiliary drives.

Project Economics – A summary of the results for the recommended system is presented in Table 15-1, Project Economic Summary. Three measures of expenditure analysis are included in the table: the internal rate of return, the savings investment ratio, and the project net present value.

A brief explanation of net benefit, savings investment ratio, and internal rate of return is as follows:

Internal Rate of Return is the cash discount rate which results in a zero net present value for the project.

Savings Investment Ratio is the present value of the discounted after-tax cash flow divided by the equity required for the project.

Net Present Value is the discount after-tax cash flow.

DESIGN OUTLINE

Steps in the design outline were development of specification, bidding, and bid evaluation of the gas turbine, followed by preparation of a ±10% cost estimate. In addition, a preliminary layout sufficiently complete to provide backup for a ±10% capital estimate was prepared. During the bidding process, turbines ranging from 2.7 to roughly 6 MW were evaluated on a life-cycle basis.

The system chosen was a combined cycle unit with a nominal electric output of 3.9 MW and a low pressure steam output of

Table 15-1

Project Description:

Capacity	4.08 MW
No-proj Gas Use	297.1 MMM Btu/Y
w/Proj Gas Use	500.7 MMM Btu/Y
Base Cost	$4.42 Million
Cost Overrun	0%

Financial Assumptions:

Year 1986 O&M	$190 Thousand
Percent Equity	100%
Interest Rate	12.0%
Years Financed	0
Corp. Tax Rate	48%

Annual Escalation Assumptions:

Cost of Nat. Gas	6.0%
Oper'tns & Maint.	6.0%
Maint. Power Rate	6.0%

Analysis Results Summary:

Equity Req.	$4.42 Million
NPV of ATCF	$7.20 Million
S/I Ratio	1.63
IRR	26.9%
Equity Payback Yr	1989

Year 1986 Operations Summary

Revenues	
Sale of Excess Power	$328.2
Expenses	
Purch. of Power (Avoided)	($2,473.3)
Maintenance Power	56.6
Standby Power	17.2
Supplemental Power	15.6
Added Purchase of Gas	1,056.2
Operations & Maintenance	190.0
Prop. Tax & Insurance	44.2
Interest	0.0
Depreciation	629.7
Cap. Int Amort	0.0
Subtotal	($463.7)
Net Income Before Taxes	$791.9

(Continued)

Financial Recap (Values in Thousands)

	1986	1987	1988	1989	1990	1991	1992	1993	1994	1995
W/o Proj. Costs										
Gas	$1,541	$1,634	$1,732	$1,836	$1,946	$2,063	$2,186	$2,317	$2,456	$2,604
Electricity	2,473	2,579	2,667	2,818	2,981	3,152	3,267	3,387	3,511	3,640
TOTAL	$1,014	$4,213	$4,399	$4,654	$4,927	$5,214	$5,453	$5,704	$5,967	$6,244
With Proj. Costs										
Gas	2,597	2,753	2,919	3,094	3,279	3,476	3,685	3,906	4,140	4,388
Electricity	89	95	100	106	112	119	126	133	140	148
Oper. & Maint.	234	248	263	279	296	313	332	352	373	396
Interest	0	0	0	0	0	0	0	0	0	0
TOTAL	$2,921	$3,096	$3,282	$3,479	$3,687	$3,908	$4,142	$4,390	$4,653	$4,932
Net Cost Savings	$1,093	$1,117	$1,117	$1,175	$1,240	$1,306	$1,311	$1,314	$1,314	$1,312
Surpl. Pwr Sales	$328	$341	$362	$385	$411	$438	$468	$500	$535	$573
Depreciation	$630	$924	$882	$882	$882	$0	$0	$0	$0	$0
Taxes (Savings)	($62)	$257	$287	$326	$369	$837	$854	$871	$888	$905
TOTAL COST SAVINGS	$854	$278	$310	$353	$400	$907	$925	$943	$962	$980

17,000 lb/hr. Nalco's Board of Directors approved the project Appropriation Request on July 26, 1984.

The natural gas combustion turbine selected is an Allison 501-KB5 unit generally referred to as a light frame aircraft derivative turbine. It consists of three main components: (1) the air compressor, (2) the combustion chamber, and (3) the power turbine. Air is compressed in the compressor section to about 200 psig, then flows through the combustion chamber where it is heated by combustion of natural gas. The hot gases are expanded in the power turbine section to provide mechanical power output. Expanded gas is exhausted from the turbine at about 1,000°F. Normal operating cycle is about 30,000 hours between major overhauls. The gas turbine-generator set is supplied by Turbosystems International. The unit will be skid mounted and is about 30-ft long. All accessories are completely installed and the unit is essentially ready for operation when it arrives at the site.

One initial concern was the noise level when the turbine is in operation. Consequently, the turbine-generator set is enclosed in an insulated cabinet to reduce the operating noise level to about 82 decibels. At this sound level, conversations can be carried on in the vicinity of the operating equipment at normal voice levels. Suitable ventilating and fire systems are installed inside the cabinet.

DETAILED DESIGN

Design included specification and purchase of remaining major components as well as issuance of plans and specifications for system installation. Two other major components were the steam boiler and the steam turbine.

Exhaust heat from the turbine is used to produce 700 psig superheated steam in the heat recovery boiler. Steam pressure is reduced to 150 psig in a back pressure turbine. The steam turbine drives a second generator which produces additional electric power.

The low pressure steam is desuperheated and put into the facility's distribution system. In summer, most of the low pressure steam is used to drive the absorption chiller to provide building cooling. In winter, steam is used for building heating.

A gas-operated duct burner is provided to produce additional steam in the heat recovery boiler for winter heating. An increase in high pressure steam production increases the electric power output from the steam turbine driven generator. Maximum system outputs on a cold day will be 5.0 MW of electric power and 35,000 lb/hr of low pressure steam.

Turbine exhaust is directed into the heat recovery boiler supplied by Energy Recovery, Inc. through insulated ducts. A gas exhaust bypass stack is located between the turbine and the boiler. This allows the turbine to continue to operate in the event of an emergency shutdown of the heat recovery boiler.

Overall system efficiency, based upon fuel input, will be in the range of 70% to 81% depending on the amount of supplemental firing. The average cost of producing electricity, after deducting an appropriate steam production cost, will be about 2.3 cents per kWh.

The gas turbine-generator and heat recovery boiler system will be controlled locally by their own microprocessor-based controllers. Overall system coordination and operating levels will be determined and controlled by Nalco's Honeywell Delta 1000 computer unit. This main unit will supervise alarms, tabulate operating data, supervise system operations, and determine the operating parameters needed to maintain optimum efficiency while satisfying the demands of the Naperville complex.

POWER PLANT LAYOUT

The power plant is being expanded by 8,000 sq ft to house the new equipment. Building expansion will cost about 1.0 million dollars; installed generation equipment about 3.9 million dollars. Equipment layout has been arranged so as to accommodate future installation of a duplicate system. This allows for potential expansion at the Naperville site through the end of the century.

When the new system goes into operation, natural gas purchases will increase and electric purchases will just about be eliminated. For instance, the current energy consumption ratio is 60% natural gas and 40% electric power.

With the new energy system in operation, the consumption ratio goes to 75% gas and 25% electricity, whereas the actual purchases go to 98% gas and 2% electricity. Excess power will be available for sale during most of the year.

UTILITY COSTS WITH COGENERATION

With the turbine and waste heat recovery system installed and the new complex occupied, the 1986 estimated net cost of utilities may be reduced to an annual rate of about 2.5 million dollars. This is a savings of 1.5 million dollars per year. The projection shows that these costs will be at about the same level in 1995 as they would have been in 1986, without cogeneration. The projected savings will result in an after-tax payback of the 4.9 million dollar investment in about 3.9 years (Figure 15-3).

These estimates are conservative because they assume constant consumption at the 1986 rate. A more likely scenario would be that Nalco will add a lab building in 1989 and another office building in 1991, and thus may increase savings. With the expected expansions Nalco may be able to justify adding a second cogeneration unit by 1993.

One important task still in progress is the negotiation of an electric power buy/sell agreement with the NED (Naperville Electric Department). NED has no generating capacity of their own. They buy all of their power from Commonwealth Edison at published rates which are under the jurisdiction of both the ICC and FERC. Any electric energy that Nalco can supply to Naperville's power grid reduces energy purchase from Commonwealth Edison. This sellback does not necessarily reduce the NED need for demand capacity.

Nalco will continue to buy supplemental power during the summer when demand exceeds capacity, maintenance power during planned shutdowns, or emergency power if Nalco experiences equipment failure. They will pay a fixed fee for administrative and system costs. From Naperville Electric's point of view, NED will pay for energy put into their system and will consider capacity purchases when available.

CONCLUSION

In conclusion, power/steam usage and the combination of existing and new facilities at Nalco's Naperville site proved to be very favorable for the installation of cogeneration. Increase in power reliability and net cost savings justified the capital investment required for cogeneration.

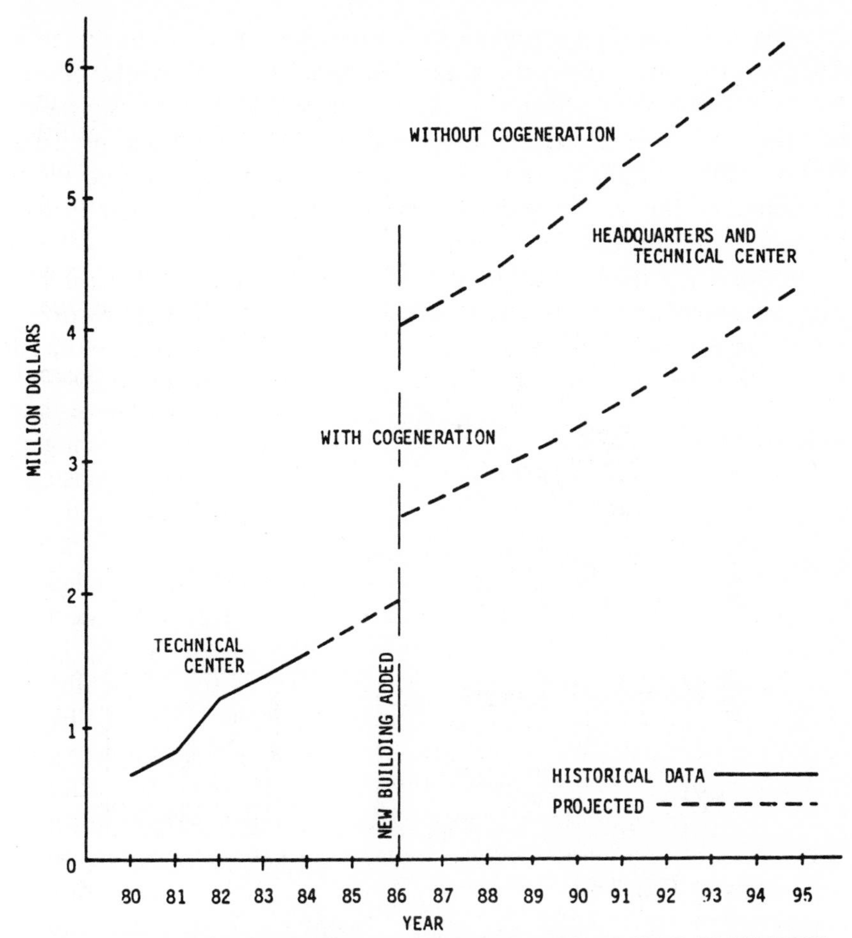

Figure 15-3. NALCO CHEMICAL COMPANY, NAPERVILLE FACILITIES ANNUAL UTILITIES COST 1986 CONFIGURATION – CONSTANT CONSUMPTION

APPENDIX A – Combined Cycle Cogeneration

This combined cycle cogeneration design incorporates a Detroit Diesel Allison 501-KB5 gas turbine, a heat recovery steam generator, and a back pressure steam turbine. The system is designed to provide 17,100 lb/hr of 150 psig saturated steam to process without supplemental firing. The condensate return from process is at about 180°F. Electric outputs from the system are 3,650 kW from the gas turbine and 270 kW from the back pressure steam turbine. Figure 15-4 illustrates major components and the approximate energy balance.

The heat recovery steam generator is a single-pressure unit designed for operation with a variable level of supplemental firing. Without supplemental firing, 17,100 lb/hr of 680°F, 700 psig steam is produced. The steam is expanded through the back pressure steam turbine to the process conditions of 150 psig saturated. Supplemental firing to an exhaust gas temperature of 1,500°F will increase steam flow to 35,000 lb/hr.

Natural gas heat input to the gas turbine averages 48.8 million Btu per hour HHV. Turbine exhaust is ducted into the heat recovery steam generator at an average temperature of 995°F with an exhaust to atmosphere from the heat recovery generator of about 350°F.

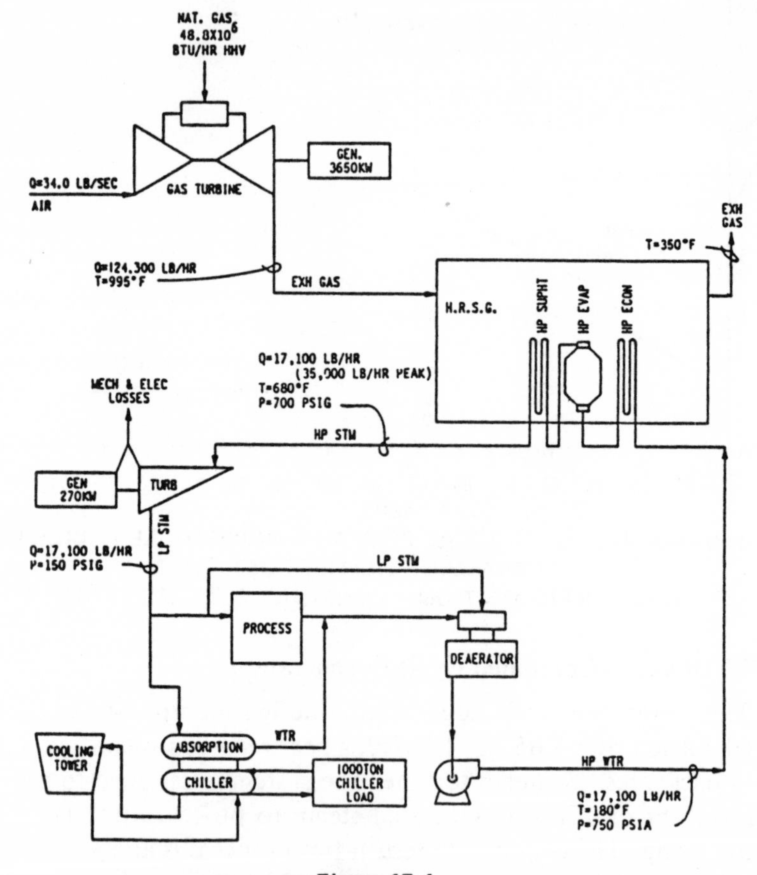

Figure 15-4
NALCO CHEMICAL COMPANY, NAPERVILLE FACILITIES SCHEMATIC DIAGRAM

PART V

COGENERATION AND UTILITIES

CHAPTER 16

A Utility-Affiliated Cogeneration Developer Perspective

Terry A. Ferrar

INTRODUCTION

For the past decade, the electric power industry has been disproportionately investing in solid-fueled generation technologies. This strategy was an appropriate reaction to the Arab oil embargo, and the associated precipitously high relative rate of inflation of oil and gas prices. Such planning has necessarily, and understandably, resulted in a rise in the cost of capacity, as capital-intensive technologies have been substituted for fuel-intensive ones.

Paradoxically, (perhaps even alarmingly), utilities and cogenerators must now reassess the market implications of a rather abrupt mitigation in relative fuel-price trends. (There are market observers that suggest that oil and gas prices may even fall in real terms over the next decade!) These new price-trend expectations have arrived just as many utilities are striving to bring capital-intensive projects into service. As such utilities are seeking (or preparing to seek) rate increases, cogenerators are positioned to offer a power supply option that is, in many applications, more consistent with today's economic environment. In effect, due to a turnaround in relative fuel-price expectations, the cogeneration industry is (in many cases) better positioned to address power market requirements.

An additional observation that requires our attention is the fact that rising prices of oil and natural gas over the past decade have tended to accelerate the development of improved technologies for efficient power production from these fuels and from nonconventional, "alternative" fuels. Small generation devices applicable to cogeneration have significantly advanced technologically. Today, the efficiency of gas turbines, combination or "packaged" cogeneration equipment, wood and waste combustion technologies, etc. provide surprisingly effective means for generating power and steam.

The combination of these events–utilities tilting their generation mix toward solid fuels, revised oil and natural gas price trends, and technology advances–permits the reasonable conclusion that the future structure of the power industry is likely to be more competitive.

You may have noticed that I have not mentioned PURPA. PURPA has had an important influence, but from the standpoint of industry development, the influence of PURPA has been vastly overstated. In any transaction "no deal is a good deal unless it's a good deal for everyone." This cliché has direct applicability to negotiations between utilities and cogenerators. The fundamental drives behind the cogeneration industry are economic–capacity position of utilities, oil and gas price trends, and technology.

In most cogeneration projects, the most economically significant transaction is between the cogenerator and the power purchasing utility. In most cases, success, or lack thereof, in negotiating this power-sale agreement, determines whether the project is realized. Today a market dichotomy exists within the utility industry that significantly influences the likelihood of a successful negotiation. Simply stated, some utilities need capacity, and others believe they do not. Let us review the negotiating environment implied by these utility capacity positions.

NEGOTIATING WITH THE UTILITY

A cogenerator approaching a utility that is in need of capacity is in the desirable position of offering a product that utility executives perceive to be valuable. Under this condition, a mutually satisfactory transaction for the sale of power may potentially be negotiated. Under such capacity conditions, utility executives feel comfortable representing to their public utility commissioners that payments for cogenerated power are appropriate–i.e., in the public interest. Such conditions permit executives to have high confidence that they will not be "stuck holding the bag" as a result of possible prudency arguments regarding their procurement of cogenerated power (in particular, the capacity portion of the payment).

Conversely, a cogenerator approaching a utility that perceives itself to have adequate capacity is not presented with a fertile negotiating environment. In such a case, attempts to use PURPA to force the utility to purchase the power (especially the capacity portion) will only result in protracted negotiations and difficult business relationships. The utility executive perceiving his company to have adequate capacity may fear that the existence of cost-effective cogenerated power will result in prudency questions regarding his chosen capacity mix—a capacity decision most likely made under different economic conditions. Thus, the cogenerator may be viewed as a threat.

The most intense negotiating environment occurs when a utility expects to bring on additional capacity, while being confronted with the availability of cost-effective cogenerated power. Under such conditions, the utility will most likely be particularly unreceptive to the cogenerator's proposals.

Simultaneously, however, the cogenerator may find many customers of the utility interested in an opportunity to avoid the rate increases expected to be associated with the utility's new generation project. Self-consumption of power by the cogeneration steam customer, or the purchase of power by a municipality or cooperative who would otherwise be purchasing from the generating utility, may be the outcome. Such an occurrence reminds utility observers of Amory Lovins' "death spiral" scenario—as the utility's fixed-costs must now be apportioned among a shrinking customer base.

The cogeneration marketplace is currently enjoying aggressive interest by power consumers attempting to avoid the rate increase (rate shock) implications associated with capital-intensive facilities entering rate base. This utility "by-pass" (as mentioned recently by John Naisbitt, the author of *Megatrends*) is perhaps the most intriguing social policy question facing the power industry today. Incidentally, this "by-pass" interest is likely to be a primary force behind policy pressure to widen power-wheeling access.

While I am discussing the changing power market, I must mention the now well-known uncertainties surrounding the definition of "avoided costs." It is interesting to note that notwithstanding the sophisticated art and science of rate-case proceedings, there still remains the question—"How much does it cost to supply an incremental unit of power?" Nevertheless, this uncertainty is real, and must be attended to in investment decision making.

It is also interesting to note that while the analytical debate goes forward on avoided costs—many utilities are finding that cogenerators can offer power *below* what the utilities believe to be their avoided costs! Indeed, quoting from this month's issue of *Cogeneration Report* "Central Main Power has signed contracts with 31 hydroelectric and cogeneration power developers—and all will receive *less than* CMP's avoided cost." Moreover, California has been forced to withdraw a "standard offer" because cogenerators were expected to produce an excess supply of power that could potentially be disruptive.

If these "less than utility avoided cost" contracts are really available from cogenerators—it leaves one to wonder what the power industry will look like in the future—i.e., what will be the cogeneration industry's share of the incremental power market?

PROSPECTS FOR THE FUTURE

Given these market observations, I would like to offer a perspective on how the character of the cogeneration industry is changing:

- Both the traditional electric power industry and the cogeneration industry are likely to face more market risks than previously anticipated – as competitive markets are more risky for suppliers, than are monopoly markets.
- As a direct result, excessively high leveraged, virtually risk-free cogeneration transactions are likely to be a thing of the past.
- It will be necessary for participants in this industry to assume the associated equity risks derived from these marketplace dynamics. Consequently, informed equity participants will be a necessity.
- As these risks are recognized a "shake-out" will occur, finding thinly capitalized, purely developer organizations receding

from the marketplace (or being limited to transactions only with utilities in need of capacity).

What should industry observers expect over the next several years?

- Some project failures, as some of the market risks I have identified begin to manifest themselves.
- Many marginal projects presently being contemplated will not succeed because of inadequate economic strength to compensate the market risks suggested above.
- As a result, the market will be smaller than many industry proponents are currently reporting.
- Financial strength or "staying power" will be a necessary ingredient for success.
- Organizations will disappear either by merger or bankruptcy.
- Industry leaders will emerge in the following areas: regional, technology based, size of facility, etc.
- Nevertheless, cogeneration will represent a significant share of incremental capacity additions in America's power market.

These observations also lead me to conclude that utility planning will increasingly need to account for these changing power-market characteristics. Effective planning for electric utilities will require recognition of the competitive nature of the power business.

The electric power business will be surprisingly dynamic over its next decade: more interesting, more risky, and hopefully more profitable for all players–cogenerators and electric utilities.

CHAPTER 17

Cogeneration Operational Issues

Malcolm Williams

INTRODUCTION

Many discussions concerning cogeneration projects focus on "avoided cost" and other legal issues which affect these projects. While these areas are extremely important and essential to the venture's success, operational issues which impact the utility and the cogenerator are equally critical. This chapter addresses the utility perspective in regard to possible impact of cogeneration systems on utility service to other customers, safety and substation operations. Other operational issues include utility transmission planning, generation planning and fuel mix decisions. All of these operational problems have an impact on the ratepayer in regard to quality of electric service and future rates. Both the cogenerator and the utility have an interest in solving these problems.

When discussing operational issues on a panel entitled *Cogeneration, A Utility Perspective,* I feel I should preface my remarks by saying that I am not speaking for the utility industry, or even my own company. I am certain, however, that these remarks may reflect some of the problems that arise in the operation of cogeneration systems from a utility viewpoint.

This chapter simply points out a few suggestions on how we look beyond the avoided cost, project financing and regulatory issues, and from a utility perspective, become aware of other issues and problems that may arise. The discussion of these problems should not be interpreted as being anti-cogeneration. In fact my company, Gulf States Utilities Company, has been in the cogeneration business for over 40 years. In Baton Rouge, Louisiana, we have been selling steam and cogenerated power to Exxon and Ethyl Corporation. Today we are selling one million pounds of process steam and the cogenerated energy that goes with it, and this 40-year-old project is still 25% more efficient than an alternative process.

Over one-half of the energy sold by my company goes to industrial customers; and almost one-half of the revenues are from industrial customers. It is essential to the long-range economic health of my company that these industries remain competitive in the world markets, operate at a profit, expand and create new jobs and opportunity along the upper Texas/Louisiana Gulf Coast. If cogeneration is a tool which can accomplish this, our company is in strong support of it. Even though my company has a large capacity margin and will probably not build additional power plants for the next decade, we would much rather see a healthy competitive cogenerating company than one which had to close.

I often hear from our many industrial customers that utilities are not consistent. They say that each utility has a different policy and that utilities are not uniform in how they handle problems. I would confess that this indictment is true. It's true because utilities are different; every utility has a different load factor, a different generating capacity reserve, different fuel mix, transmission system, and many other factors which cause each utility to view cogeneration in a different light. On the other hand, I would say that most industries are different. We deal almost on a daily basis with cogenerators who want to burn gas or oil, coke, coal, wood chips, cow chips, rice hulls, methane, garbage and any other matter that will burn in the boiler. These projects range from firm offers of 400 megawatts to 800 kW once a week. Almost without exception each one of these potential cogenerators says that he is different, he must operate in a special way, he must have a different kind of contract, different interconnection, and a different avoided cost. The point is that both industry and utility have unique problems, and that the participants in these cogeneration projects must take some time to understand the other's concern. The utility generally is concerned with how to operate a system and maintain reliable service without impacting ratepayers or stockholders. The cogenerator is generally concerned about making the system work, being able to deliver the energy and provide the rate of return which encouraged him to start the project in the first place.

OPERATIONAL PROBLEMS

The subject of this chapter however, is operational problems, which can be summarized in several areas. The first set of problems comes from the concerns of utility system operators. Utility systems have been, and are being designed by electrical engineers who have adopted operating standards, public safety standards, work procedures for lineman and specifications for providing service. These system operators probably feel how AT&T initially felt when MCI and Sprint decided to tie on their system. Utility operators have a genuine concern that cogenerators on their system may impair the quality of service to other customers. They have a concern over voltage, flicker, harmonics, operating continuity and power factor. The cogenerator who suddenly finds himself in the electric business is sometimes slow to adapt to utility methods and work practices. Right or wrong, it's a problem for the utilities.

The next set of utility concerns comes from the generation planners. PURPA was passed using the utility economics and transmission system to make it work, but nobody asked the utilities. The generation planners who forecast, plan and construct generation plants get frustrated when somebody takes over that planning for them. They become a reactive group to the cogenerator who decides on a cogeneration project, and through regulatory action the utility then must defer their generation plans to accommodate them. So what the generation planner works on and once calculated as the needs of the utility in regard to location, in-service date and reliability may or may not match up with potential co-generator plans. Another utility problem.

Problems for utilities also come from the fuel procurement people or those responsible for fuel planning. Most utilities, at least in recent years, have plans to diversify their fuel mix. Particularly those utilities along the Gulf Coast that planned to get off of natural gas and convert to coal, lignite and nuclear fuels for the long term. In a cogeneration project, the cogenerator, not the utility, determines what fuel that he will use, and in some cases where a generating plant is deferred, the utility must accept a gas based cogeneration system in lieu of lignite for instance. The fuel planner feels that this will be an increased reliability risk in a few years when vanishing gas bubbles or rising gas prices put the ratepayer on a gas-based economy in contrast to the fuel planner's strategy of being on lignite or coal. This short-term compromise of a long-range goal is another utility concern.

Another utility problem concerns transmission planning. Utility engineers and transmission planners have designed the system to meet the utility loads, generation and import power; and they effectively build that system with the most practical engineering manner to meet system needs. Within a utility system, however, this planning becomes more complex when a cogenerator decides on the location to build a plant. It would certainly be an oversimplification to assume that the existing utility substation is adequate for the cogenerator to tie in, as is, and sell energy. For example, utilities have designed the power flow in one direction and the cogenerator needs are different. The problem on the utility system may be aggravated to the extent that breakers, relays, lightning arrestors, metering and other equipment must be replaced not only for the cogenerator interconnection, but for other customers as well. In some states, too many plants in one area have proven that the transmission system is no longer adequate. There are instances in which utilities either now or in the future will be forced to spend millions of dollars to upgrade transmission facilities to accommodate cogenerators. This not only presents an engineering problem, but a regulatory problem in the sense that someone must determine who pays the bill. Is it the ratepayer, the utility, or the cogenerator?

Utilities also have a regulatory problem. This regulatory problem impacts industry planning as well as utility planning. Many states still do not have a final resolution on such items as avoided cost, standby power, backup rates and interruptible rates, all of which make it difficult for a utility to determine what their obligations will be in the future. For instance, will there be a limitation on backup power? Will the utility be required to stand by to 10% of the load or 110% of the load? Will the regulations be changed in the future? All good questions, but no immediate answers, which presents problems for the utility.

The case is probably overstated, but only to show that there is an operational side of cogeneration which creates problems. I have given you those problems from a utility perspective, only to emphasize that in good workable cogeneration projects we must look beyond the contractual commitments and financial stability of the project, and assure ourselves that these operational problems are minimized both for the utility and the industry. I think that both the utility and industry share a concern for other non-cogenerating customers and the protection of the ratepayer.

Most of the operating problems discussed in this chapter can be solved by good communications and by utility and industrial operating people. I find that operating people and engineers are practical, and almost without exception can work out the difficulties. In negotiating cogeneration contracts, however, I cannot necessarily make this statement for lawyers, financial people, or accountants.

Operating problems are not unique to cogeneration. Any project or business transaction has problems as well as benefits. In cogeneration, both the utility and industry agree that a project that benefits one of the parties at the expense of the other will be unsuccessful. Successful projects are those in which industry and utility cooperate and work towards resolving the problems. Projects where everyone wins will be those that we are most assured will have the greatest long-term benefit on both industry, the utility, the ratepayer and the industry and utility stockholders.

CHAPTER 18

Negotiating A Favorable Cogeneration Contract With Your Utility Company

Don H. Lark and Jack Flynn

INTRODUCTION

A relatively small cogenerator may find it difficult to negotiate a favorable cogeneration contract with a relatively large utility. This chapter will tell prospective cogenerators how to make sure the contract they negotiate meets their energy needs, while achieving their financial objectives.

PART I

From the Cogenerator's Point of View – *by Don Lark*

There are two sides to any story. Therefore, I will tell this story from the cogenerator's viewpoint and my colleague, Mr. Jack Flynn of Utility Savings Unlimited will tell it from the utility's side.

Negotiating a cogeneration contract with the utility company is unique in the sense that you are a competitor in the power production business. In most cases, the utility neither needs nor wants your power. However, Federal law says that if you meet certain conditions, the utility must buy your power.

What are these conditions?

- You must file a "Notice of Application for Commission Certification of Qualifying Status as a Cogeneration Facility."
- You should write a "Letter of Intent" to your local utility informing them of your intention to install a cogeneration facility.

Performing these two acts now places you in a legal position to talk with your utility. It also obligates the utility to respond to you.

From this point on, the benefits you negotiate into the contract are in proportion to the amount of homework you do.

Homework consists of:

- Obtaining the "Standard Offers for Power Purchases from Qualifying Facilities" from your local utility. These come in several versions.

  a. Firm Capacity

  b. As Available

  c. Wheeling to Another Utility

  d. Small Power Generation Under 100 kW

- Contacting your Public Utility Commission or whatever other body regulates utilities in your state. Learn from them what rules and regulations they impose on the utility in regard to cogeneration contracts.

- Hiring an attorney versed in dealing with utilities and cogeneration. Have your attorney review the "Standard Offer" from your utility. Also inform your attorney of your cogeneration requirements and operating conditions. Then, if you feel you can improve on the agreement in your favor, have your attorney prepare a counter offer to present to the utility.

Before you can deal from a strong base, you must know the rules of the game and how it is played.

At a cogeneration project of mine in California that produces 21 megawatts, we were able to negotiate a very favorable contract with the utility. It is still a one-of-a-kind contract for that area.

The contract contained a "floor," or a bottom line price, which the utility could pay for the purchase of the cogenerator's electricity. In this case, the lowest price the utility can pay is 6¼¢ per kWh. This provided the cogenerator with reasonable assurance that the investment would pay out in 2½ years. However to get this agreement, we had to give up something. We agreed to accept 90% of the avoided cost payment whenever it is above 6¼¢.

Contract negotiations began in July, 1980 and were finally agreed upon in August, 1981, after many hours of talks and redrafts. As you can see, negotiations can be protracted.

On this project, we did our homework by studying all the rules and regulations, filing the necessary documents, and utilizing not only the company attorney, but also an outside legal firm that specialized in utility negotiations. We obtained the PUC guidelines, prepared a proposal, and presented it to the local utility. The utility responded by providing rate sheets and its preliminary proposal for interconnection.

Of the various power purchase contracts offered by the utility, we chose the "As Available" contract. A "Capacity" contract obligates the cogenerator to make a specific amount of power available to the utility over a period of years. The penalty for this type of contract was too risky for our operation. Whereas the "As Available" contract, which offers a lower price per kWh, does not carry the obligation to deliver or the excessive penalty clause for non-delivery of power.

The utility proposal contained a standby charge of $1 per kWh for power availability if the cogeneration facility failed. This was not acceptable to us and we were able to negotiate it out of the contract. However, if power from the utility is needed, a demand charge would apply.

The 21-MW cogeneration facility produced all the steam and electrical needs for the plant–180,000 lb. of steam per hour and 6-8 MW of power, with the remaining 13-15 MW sold to the utility.

To sell power to the utility, the cogeneration system is connected in parallel with the utility at all times. Parallel operation has the advantage of reducing the required amount of coordination between the cogenerator and the utility, but it has the disadvantage of subjecting both systems to disturbances occurring in either.

Selection of the interconnection point and equipment is also a major factor in negotiating with the utility.

The main switch station for the California project was 12 KV and located about 1,600 feet from the property line. Service delivery from the utility to this main station was by overhead power lines, which, by the way, have always been a nuisance when operating cranes at the plant.

The utility proposal for the interconnection was a 69-KV substation. It was to be located on 7,500 sq. ft. of company property, would cost more than $1 million and have a fixed annual maintenance fee of $40,000. We countered with a proposal to bring our own power lines up to the property line to avoid a penalty for line losses and to tie directly into the utility's 12-KV bus, which is just across the fence. The utility has a power generating plant at this location.

We obtained the utility's one-line for the bus and switchyard. My consulting electrical engineer determined that all that was needed was a reactor to limit groundfault inrush to 250 MVA. This was presented to the utility and finally agreed upon at a very reduced cost of $102,000–a far cry from the original $1 million proposed by the utility. My consultant paid for himself many times over.

The utility wanted to design and install the interconnect facility. However, as long as we were paying for the installation, we wanted control of the equipment type and manufacturer. Negotiations resolved that we would design, purchase and install the interconnect facility on company property. The utility need only install the reactor on their property.

We did this for about 30% less than the utility would have spent for the same thing. With this control, we were able to coordinate the overall equipment as to type and manufacturer to meet company standards. However, the installation and equipment had to be approved by the utility to meet their standards for the required protection equipment for both systems.

This contract has been used as a model for power purchase agreements in other areas. It is still one of a kind in San Diego, but Southern California Edison in Los Angeles and Pacific Gas & Electric in San Francisco have several agreements of this type.

PART II

From the Utility's Perspective – *By Jack Flynn*

Utilities have had monopoly control over their energy products for many years. Recently, market forces have threatened that control and are now moving many utilities toward a new strategy of marketing diversification that emphasizes energy services rather than energy products.

Most utilities are now moving along a path that I call the "Utility Diversification Timeline." "Control" is at the near end of that timeline. "Diversification" is at the far end.

Some of the more traditional electric utilities are determined to stay at the near end of the timeline in the "Control" stage—or what I call the "Reddy Kilowatt" stage. Although we have not seen much of him lately, Reddy Kilowatt was the electric utility industry's highly popular advertising mascot during their hard-selling advertising campaigns back between the end of World War II and the beginning of the Age of OPEC.

Other more forward-looking utilities like SDG&E are moving on down the timeline toward the diversification end—or what I call the "ESCO" stage. An electric utility that becomes an ESCO—or "Energy Service Company"—no longer serves as just a hardseller of electricity. As an ESCO it assumes a new role as designer and marketer of energy services that meet their customers' special energy needs.

After a somewhat sluggish start, SDG&E is now on the move toward diversification. A review of the events that led to their decision is illuminating:

- 1950's and '60's . . . electric sales increase rapidly in San Diego, one of the five fastest growing areas in the U.S.
- Early 1970's . . . rapid escalation of electric rates by San Diego's primarily oil and gas-fired utility results from the Arab Oil Embargo and subsequent OPEC market manipulations.
- Late 1970's . . . efforts to get SDG&E off foreign oil by developing a coal-fired power plant in Utah and a nuclear plant in the California desert failed.
- 1980 . . . in a debate with decentralized power advocates, Amory Lovins and Daniel Yergins, SDG&E board chairman Bob Morris argues for more large central power plants.
- 1981 . . . under CPUC pressure, SDG&E reluctantly begins a promotion program for customer cogeneration.
- 1983 . . . the CPUC threatens to fine SDG&E shareholders for "foot dragging" in the promotion of cogeneration in their service territory.

- 1984 . . . Bob Morris leaves SDG&E to head Energy Factors, a major cogeneration marketing and management company that SDG&E had developed and then sold off.
- 1985 . . . SDG&E's new board chariman Tom Page announces a new diversification strategy that contemplates utility financing of customer cogeneration projects and marketing of management and maintenance contracts for customer-owned cogeneration plants.

In the early 1980's, the majority of SDG&E's power was generated by their own fossil fuel-fired plants. Now, the majority of their power is purchased from outside sources, including cogenerators. As they lost control over their power supply, SDG&E's attitude toward cogeneration changed. Early on, they dragged their feet in promoting cogeneration in their service territory. Now, they are even considering financing local cogeneration projects.

How will your local utility react to your proposed cogeneration project? That probably depends where they stand on the "Utility Diversification Timeline."

If your local utility is still in the "Reddy Kilowatt" stage, their primary concern will probably be to maintain control over their energy supply. They will want an ironclad capacity contract with strong penalties for your failure to deliver. You will have to impress them favorably with your project's long-term stability.

If your local utility is turning into an ESCO, talk to them about your cogeneration project as an investment opportunity. Maybe they will want to finance it for you, or even operate and maintain it for you on a contract.

Index

G

H

I

J

K

L

M